CONTENTS

Hawai'i is blessed with an abundance of rain. Numerous waterfalls cascade down into a lush mountain forest and continue on to enrich the wetlands below. (JIM DENNY)

PREFACE

This book is for those who want to identify the many birds they encounter in their travels throughout the Hawaiian Islands. It includes nearly every species of bird on land and at sea in the main Hawaiian Islands. In total, 170 species or subspecies are described and illustrated. It does not include species thought to be extinct or those species found only in the inaccessible Northwestern Hawaiian Islands. Also excluded are alien and migrant species for which there are only one or two records.

Not all the birds contained in this guide are common. It should be noted that many of the species illustrated are rare and will be seen only by the persistent observer. This is especially true for migrant wetland birds. The same can be said for endangered forest birds. Densities of endangered native forest birds are greater in more remote, difficult-to-reach areas. Some species are not likely to be seen without special permit for access to private land or to enter restricted habitat. Possible locations are given for all species where applicable.

The scientific and English names used in this book are derived from the *Checklist of the Birds of Hawaii – 2002* as compiled by Robert L. Pyle. The only departure from Dr. Pyle's list is the addition of a possessive apostrophe to the names of some species. A sightings database, also compiled by Dr. Pyle, was used to determine the frequency with which a species occurs in Hawai'i. Additional information was gleaned from an archive of sighting reports submitted by members of the Hawai'i Birding Yahoo Group. This archive contains sightings spanning ten years as managed by Christian Melgar on his excellent Birding Hawaii Web site: http://www.birdinghawaii.co.uk. A table detailing this combined information can be found in Appendix 1. The reader is cautioned that the rating a species is given (rare, uncommon, common, or abundant) can be misleading. A species can be common at the right time of year in a specific habitat and rare everywhere else at all other times. The rating given reflects the best of circumstances.

The Hawaiian names for the birds and plants as they appear here incorporate the pronunciation guides used in the *Hawaiian Dictionary* by Mary Kawena Pukui and Samuel H. Elbert (1986). Locations are spelled as described in *Place Names of Hawaii* by Pukui, Elbert, and Esther T. Mookini (1974). The *kahakō* or macron (¯) denotes a stressed vowel, and the *'okina* (') marks a glottal stop. English names for Hawaiian birds are listed where possible, and a scientific name is given for every species. Those species designated as "endangered" or "threatened" by the U.S. Fish and Wildlife Service are noted. Each species is also labeled as to its status in Hawai'i: alien, migrant, indigenous, or endemic. Nearly all measurements are from *A Field Guide to the Birds of Hawaii and the Tropical Pacific* by H. Douglas Pratt, Phillip L. Bruner, and Delwyn G. Berrett (1987). For measurements not available in that text, the source consulted was *Complete Birds of North America* by The National Geographic Society (2006).

Although the bulk of the information in this text is from my own observations, also included is a substantial amount of material collected by others. The serious observer or naturalist who desires more comprehensive data about the flora and fauna of Hawai'i can find a wealth of information among the references listed at the end of this book.

Hawai'i's birds can be separated into various categories. In this text I have elected to assemble them into five basic groups: urban birds, country birds, forest birds, wetland birds, and seabirds. For some species there is substantial overlap between habitats.

I encourage the reader to write in the margins of this book: date seen, location, behavior, or any other fact to help you remember the experience for years to come. In your travels in Hawai'i, if you happen upon a species not listed in this book, please e-mail me (jimdenny@hawaii.rr.com or kauaibirds.com). Good birding!

ACKNOWLEDGMENTS

This book would not have been possible without the help of others.

Nearly a third of the photos in this book are the work of noted, award-winning wildlife photographer Jack Jeffrey. Furthermore, he can be partially credited for many of my own photos because he has been a hiking companion on many photographic adventures throughout the Hawaiian Islands during which his guidance in locating birds and his suggestions for getting the best images were invaluable. More of his highly acclaimed images can be found on his Web site, jackjeffreyphoto.com.

Some of the seabird photos are here only because of Sharon Reilly of the Save Our Shearwaters program on Kaua'i. Sharon would often notify me of seabirds that were turned in to her organization. She graciously allowed me to photograph them upon release.

Michael Walther, of Oahu Nature Tours, provided me with several hard-to-get images. More of his wonderful photos can be found at http://www.oahunaturetours.com.

Additional images were provided by George L. Armistead (http://www.fieldguides.com), Mike Bowles and Loretta Erickson (http://www.amazornia.us), Gerhard Hofmann (http://www.hofmann-photography.de/html/welcome.html), Peter LaTourette (http://www.birdphotography.com/), and Alan and Elaine Wilson (http://www.naturespicsonline.com). Their Web sites give a glimpse into the abilities of these gifted photographers.

Dr. Robert Pyle of the Bishop Museum and his son Richard provided me with the historical data of bird sightings in Hawai'i. Bob Pyle passed away as this book was in preparation. He will be greatly missed by the Hawaiian birding community.

A huge debt of gratitude goes to Christian Melgar, a London, United Kingdom, resident, who created and maintains an excellent Web site devoted to birds in Hawai'i: http://www.birdinghawaii.co.uk. The information he includes on his Web site was extremely helpful in the preparation of this book.

I am also in debt to all those who took the time to notify me of the location of a hard-to-find species or who aided in the preparation of this book. Among them are Abby Brown-Watson, Sheila Conant, Marguerite Crothers, Eileen D'Araujo, Reginald David, Glenn Denny, Arleone Dibben-Young, Matthew Fletcher, Eric Isley, David Kuhn, Keith Leber, Pauline Roberts, David Watson, Brenda Zaun, and the members of the Hawai'i Birding Yahoo Group.

Warmest mahalo.

AN INTRODUCTION TO HAWAI'I'S BIRDS

Serious birders often include Hawai'i in their list of "must-go" places because these Islands are home to a wonderful assortment of birds, some of which can be seen nowhere else. In addition to a beautiful variety of native forest and native wetland species are graceful seabirds, long distance migrants, and an abundance of introduced birds from all over the world. More than 300 species of birds have been recorded in the Hawaiian Islands and the number is still growing. Almost every year another species is added to the list, usually a new migrant or unauthorized introduction. You just never know what will turn up.

The most sought-after species, of course, are Hawai'i's endemic forest birds. Hawai'i's isolation in the middle of a vast ocean allowed a degree of adaptive radiation exceeding that of any other island group. The bones of long-extinct birds still emerging from fossil deposits and ancient lava tubes continue to dazzle researchers with the extreme variety of these adaptations. Scientists believe that from a small flock of finches, perhaps blown here by a hurricane, evolved more than 50

The Small Indian Mongoose (Herpestes auropunctatus) *was introduced to the Islands in 1883 to combat rats in the sugarcane fields. It has become an efficient predator of island birds, particularly ground-nesting species. It is present on all the islands except Kaua'i and Lāna'i.* (JACK JEFFREY)

The Black Rat (Rattus rattus) *is an agile climber that preys upon the eggs and nestlings of birds.* (JACK JEFFREY)

Wild pigs do tremendous damage to the native forest of Hawai'i. They knock over native ferns to eat the starchy inner core and dig up the forest floor in search of roots and worms. The soil they disturb in the process then becomes fertile ground for invasive alien weeds and stagnant breeding pools for mosquitoes. (JACK JEFFREY)

separate species and subspecies. Two hundred years ago, when naturalists began to collect them, they did not seem to fit into any established taxonomic group. A whole new subfamily had to be added: Drepanidinae. Informally, they named them honeycreepers. Unfortunately over half are now extinct, the victims of introduced diseases and predation. The story is not yet over. Avian malaria, avian pox, feral cats, feral dogs, feral pigs, a mongoose, and three species of rats continue to contribute to the decline of this precious resource. There is still much to be seen, however. At this writing, the magnificent long-billed *'I'iwi* is still relatively common in the upland native forests of Kaua'i, Maui, and the Big Island of Hawai'i.

Seabirds are another reason birders are eager to come to the Hawaiian Islands. These birds are indeed inspiring to watch as they glide so gracefully on the refreshing trade winds. By far, the best and the easiest place to see them up close is the Kīlauea Point National Wildlife Refuge on the island of Kaua'i. Eight species of seabirds regularly visit this historic lighthouse site. Here too, surprises occasionally occur. A Kermadec Petrel and White-tailed Eagle have both been reported there.

The well-maintained West Loch Shoreline Park near Pearl Harbor is an excellent place to see a mix of urban birds and migrant shorebirds. (JIM DENNY)

Avian pox is a serious threat to the survival of Hawai'i's endemic forest birds. Mosquitoes act as a vector for this viral disease. When birds are bitten on the feet or around the eye, disfiguring lumps and blindness can occur. Badly infected birds like this 'Apapane *lose the ability to perch. Unable to forage for food, they fall to the ground to die of starvation or become easy prey for rats and feral cats.* (JIM DENNY)

Native wetland birds and migrants contribute the largest number of species to the list of birds recorded in the Hawaiian Islands. It is hard to believe that a migrating shorebird can depart Alaska or Siberia and find a place as small as Hawai'i, a mere speck in a huge ocean, but it happens annually in large numbers. In the winter months, the Pacific Golden-Plover is a common sight on lawns and in parks in Hawai'i. The Wandering Tattler, the Sanderling, and the Ruddy Turnstone are also among those that regularly make the nonstop 2,500-mile (5,025-km) journey.

Introduced or alien birds also constitute a large portion of Hawai'i's birdlife. In this group are some of the most abundant and conspicuous birds in the Islands. Most are intentional introductions. Some are escapees from cages. Many countries and continents are represented—a dove from Asia, a myna from India, a cardinal from South America, a skylark from Europe, and a quail from North America, to name a few.

There is enough to occupy even the most ardent birder.

Kapi'olani Park, the grounds of the Honolulu Zoo, and the congested walkways of Waikīkī are home to several species of urban birds, some of which cannot be seen on other islands. (JIM DENNY)

URBAN BIRDS

NEARLY EVERY BIRD now found in our lowland parks, towns, and cities is an alien species. Many were intentionally introduced as a substitute for declining native birds. Some were selected for their beauty or for their song. Others were imported in an attempt to control insect pests in agriculture. Some are, themselves, serious agricultural pests. A growing number are escapees or deliberate releases from the pet trade. Some are so widespread that they can be found on every island in almost any habitat. Others have a more limited distribution. Nevertheless, the reader is sure to recognize most of these birds. We encounter them daily during our outdoor activities.

Rose-ringed Parakeet

PSITTACULA KRAMERI

Alien
16 inches (41 cm)

This escaped cage bird is the first parrot to become well established in Hawai'i. Native to central Africa and India, the Rose-ringed Parakeet can currently be observed with relative ease only on O'ahu and Kaua'i. A few reports have also come from Maui and the Big Island. It is an attractive species with a long blue green tail and striking red bill. The Rose-ringed Parakeet is a serious agricultural pest and can do extensive damage to seed, fruit, and vegetable crops. It will eat corn, lychee, papaya, and even bell pepper from the backyard garden. At night, this noisy bird prefers to roost in Royal Palms. On O'ahu, birds can usually be seen near the archery range in Kapi'olani Park, at Foster Botanic Garden, and on the University of Hawai'i Mānoa campus. On Kaua'i look for the large flocks that fly out of Hanapēpē Valley around sunset or those that roost on the grounds of the County Building in downtown Līhu'e.

Rose-ringed Parakeet, Psittacula krameri. (JACK JEFFREY)

Red-crowned Parrot

AMAZONA VIRIDIGENALIS

Alien
13 inches (33 cm)

The Red-crowned Parrot, or Red-crowned Amazon, is native to Mexico. The free-ranging population established in the wild on O'ahu is due to escaped or intentionally released birds. Adult males are dark green above and pale green below with a scaled appearance. The crown and forehead are red. The red is limited to the forehead in females. A narrow whitish ring surrounds the eye. The nape is bluish green, and the bill pale yellow. The tips of the primaries are black. It has a shorter tail than the similar-sized Red-masked Parakeet. The Red-crowned Parrot has a loud, harsh call that is often uttered in flight. This noisy species favors lowland and urban areas, where it feeds on fruit, nuts, and flower buds. A flock of a hundred or more birds is known to roost in upper Waimano Valley. Birds have been seen in Mililani, Waipahu, and Pearl City.

Red-crowned Parrot, Amazona viridigenalis. (BOWLES/ERICKSON, AMAZORNIA.US)

Red-masked Parakeet

ARATINGA ERYTHROGENYS

Alien
13 inches (33 cm)

The Red-masked Parakeet, or Red-masked Conure, is native to Ecuador and Peru. It is an attractive green bird with a long, pointed tail. The crown, face, and shoulders are red. The bill is cream colored, and a conspicuous, wide white ring encircles the eye. Legs and feet are gray. Red-masked Parakeets are noisy opportunists that feed on a large variety of fruits and seeds. They often fly great distances in search of food. A large flock of fifty or more of these birds lives in the vicinity of Diamond Head Crater on O'ahu. This species is also established in the Kailua-Kona area of the Big Island where they can often be found in the *kiawe* trees behind the Keauhou Small Boat Harbor. Reports of other parrots in Hawai'i are increasing. The Mitred Parakeet *(Aratinga mitrata)* and Patagonian Parakeet *(Cyanoliseus patagonus)* are just two more of nearly a dozen species known to be breeding in the Islands.

Red-masked Parakeet, Aratinga erythrogenys. (JACK JEFFREY)

Japanese White-eye (Mejiro)

ZOSTEROPS JAPONICUS

Alien
4 inches (10 cm)

The Mejiro, as it is known in Japan, was introduced from that country in 1927. It has adapted extremely well to Hawai'i and can now be found from the wettest summit to the driest plain on all islands. Part of their success may be due to their ability to exist on a variety of food sources including fruits, insects, and nectar. Adult males are green with a yellow throat and white abdomen. Females are somewhat duller. The diagnostic feature that quickly separates the adult from all other small, green forest birds is the prominent white eye ring. The Japanese White-eye travels in small twittering flocks that become noisier when disturbed.

Japanese White-eye (Mejiro), Zosterops japonicus. (JIM DENNY)

Spotted Dove

STREPTOPELIA CHINENSIS

Alien
12 inches (31 cm)

The Spotted Dove is native to India and Southeast Asia. It was introduced to Hawai'i in the 1800s. Other names for this species are Lace-necked Dove or Chinese Dove. It is a common sight in Hawai'i in both rural and populated areas and can be easily approached in areas high in human traffic such as parking lots and city sidewalks. The Spotted Dove is much larger than the equally common Zebra Dove. Adults are gray or grayish brown. A conspicuous black patch on the back of the neck is spotted with white. Other features are a red iris and red legs and feet.

Spotted Dove, Streptopelia chinensis. (JIM DENNY)

Rock Pigeon

COLUMBA LIVIA

Alien
12 inches (30 cm)

The Rock Pigeon, also known as the Common or Domesticated Pigeon, was one of the first birds introduced to Hawai'i after European "discovery." It was imported from Europe about 1796. Nearly all populations of this species are closely associated with humans. In urban areas they roost on telephone and power lines or on the ledges of buildings. They often can be seen circling over populated areas in large flocks. Rock Pigeons express considerable variety in the color of their plumage, although there is a preponderance of white in the Honolulu population.

Rock Pigeon, Columba livia. (JIM DENNY)

Zebra Dove

GEOPELIA STRIATA

Alien
8 inches (20 cm)

The Barred Dove, as it is sometimes called, is one of Hawai'i's most abundant lowland birds. The repetitive staccato call of this small dove is a major part of the morning sounds of the Islands. Like the Spotted Dove, the Zebra Dove is equally common in both rural and urban areas. In our cities and towns, they can become quite tame almost to the point of being underfoot. The Zebra Dove gets it name from the many striations prominent on the neck and breast. It varies from gray to light brown and has light blue facial skin. This small dove was imported from Australia in 1922.

Zebra Dove, Geopelia striata. (JACK JEFFREY)

Mourning Dove

ZENAIDA MACROURA

Alien
12 inches (31 cm)

This is Hawai'i's least common dove. All reported sightings of this species have previously come from O'ahu, Maui, and the Big Island, but the bird has recently been sighted on Moloka'i as well. It is about the same size as the Spotted Dove but more slender with a long, pointed tail. The back is grayish brown punctuated with several black spots. The head and neck are a lighter gray with a pinkish wash. There is a small black crescent below the eye. In flight, its wings make a quivering whistle. The Mourning Dove is a recent (1964) introduction to Hawai'i from North America, where it is extremely common. The name comes from the mournful call of the bird.

Mourning Dove, Zenaida macroura. (JACK JEFFREY)

Red-vented Bulbul

PYCNONOTUS CAFER

Alien
8.5 inches (21 cm)

The Red-vented Bulbul is established only on the island of O'ahu. There have, however, been rare reports of it from other islands. The unfortunate presence of this species in Hawai'i is probably the result of an unauthorized release in 1966. The head is black with a small crest. The vent is red and the rump white. There is a band of white at the tip of the tail. Like the Red-whiskered Bulbul, this tropical Asian species is a serious agricultural pest. It not only eats fruit but devours the buds of orchids and hibiscus as well. Sightings of this species on islands other than O'ahu should be reported to the State Department of Land and Natural Resources.

Red-vented Bulbul, Pycnonotus cafer. (JIM DENNY)

Red-whiskered Bulbul

PYCNONOTUS JOCOSUS

Alien
7 inches (18 cm)

Currently this tropical Asian species is established only on the island of O'ahu. It is a fairly recent (1965) introduction, presumably the result of a released or escaped cage bird. A red patch, or whiskers, on the face gives this attractive species its name. There is also an area of red at the vent. Another prominent feature is the conspicuous black crest. Although not as plentiful or as destructive as the Red-vented Bulbul, the Red-whiskered Bulbul is still a serious agricultural pest. Sightings of this species on islands other than O'ahu should be reported to the State Department of Land and Natural Resources.

Red-whiskered Bulbul, Pycnonotus jocosus. (JIM DENNY)

Common Myna

ACRIDOTHERES TRISTIS

Alien
10 inches (25 cm)

This ubiquitous species has been here for generations and is Hawai'i's most easily recognized bird. Introduced in 1865 from India in an attempt to control armyworms in pasturelands, the myna is now common and widespread in Hawai'i. It is a very social bird, traveling in pairs or small groups. The myna does not hop like cardinals or finches. It prefers to walk or skip instead. Occasionally, they gather in parks by the dozens for posturing exhibitions or what local children describe as "myna bird fights." Mynas also have the curious habit of using their bill to pick up and display a large piece of shiny plastic or paper. The Common Myna is brown and black. White wing patches are conspicuous in flight. The legs, bill, and skin behind the eye are yellow.

Common Myna, Acridotheres tristis. (JIM DENNY)

Northern Mockingbird

MIMUS POLYGLOTTOS

Alien
10 inches (25 cm)

The Northern Mockingbird, introduced in 1928 from North America, is common in dry habitats throughout Hawai'i. In the spring and summer months this species often seeks a high perch from which it can sing its territorial song. As the name suggests, the Northern Mockingbird is a great imitator. Its repertoire includes the songs of other birds and even the "meow" of a cat. It actively and noisily defends its nest territory by dive-bombing the intruder. On moonlit nights, the Northern Mockingbird can occasionally be heard singing late into the evening. Adults are gray above and pale gray below. The tail is long and black. When feeding on the ground, the Northern Mockingbird often methodically opens and closes its wings revealing two bold white wing bars.

Northern Mockingbird, Mimus polyglottos. (JIM DENNY)

Northern Cardinal

CARDINALIS CARDINALIS

Alien
9 inches (23 cm)

This attractive species, also known as the Kentucky Cardinal or Virginia Cardinal, can be found throughout the Islands. It is readily attracted to backyard feeders, especially those containing sunflower seeds. Although the Northern Cardinal is present in the wet native forest, its numbers are much greater in the lowlands. In the morning and evening hours of spring and summer it can frequently be heard singing from high perches. Males are bright red and females a dull reddish brown. This bird was imported from North America in the late 1920s.

Northern Cardinal, Cardinalis cardinalis. (JIM DENNY)

Red-crested Cardinal

PAROARIA CORONATA

Alien
7.5 inches (19 cm)

Often called the Brazilian Cardinal, this species is one of Hawai'i's most beautiful birds. It is present on all islands except the Big Island and is a common sight on hotel grounds, golf courses, parks, and even city streets, where it forages in pairs or small family groups. In areas high in human traffic it can be quite tame. Adult birds are gray above and pure white below. The head is a striking bright red with an erect crest. Juveniles differ in having a brown head and dark bill. It was imported from South America in the 1930s.

Red-crested Cardinal, Paroaria coronata. (JIM DENNY)

Yellow-billed Cardinal

PAROARIA CAPITATA

Alien
7 inches (18 cm)

The Yellow-billed Cardinal, a South American native, was introduced to Hawai'i around 1930. It is bluish black above and white below with a black throat. The head is bright red, and the bill and legs are yellow. This species resides only on the Big Island, where it is most numerous on the leeward coast. Populations are rapidly expanding, however, and birds can now also be seen without too much effort in the Hilo area. The similar Red-crested Cardinal, a species not yet established on the Big Island, is grayer on the back, all white below, and has a prominent crest.

Yellow-billed Cardinal, Paroaria capitata. (JACK JEFFREY)

House Sparrow

PASSER DOMESTICUS

Alien
6 inches (15 cm)

The House Sparrow is present on all the islands of Hawai'i. It is a common bird near human habitations. It often can be seen around schoolyards, beach parks, picnic tables, and outdoor restaurants: places where it can feed on the crumbs humans leave behind. It is occasionally found in agricultural areas but rarely in the forests. Some observers contend that the Hawaiian population is not as boldly colored and has paler legs than continental birds. The House Sparrow is native to Eurasia but was imported to Hawai'i from New Zealand in the early 1870s.

House Sparrow, Passer domesticus. (JACK JEFFREY)

House Finch

CARPODACUS MEXICANUS

Alien
6 inches (15 cm)

The House Finch, often called the Papaya Bird (because of its fondness for overripe papaya fruit), is one of the most common birds in Hawai'i. It is seen in large numbers in urban and agricultural areas on all the Islands. It can also occasionally be seen in small family groups within the native forest. There is considerable variation in the breeding plumage of male House Finches in Hawai'i. Some birds are predominately yellow on the breast, head, and rump, but others are mostly orange red. Female House Finches are brownish and heavily streaked below. This species was imported from North America in 1869.

House Finch, Carpodacus mexicanus. (JIM DENNY)

Saffron Finch

SICALIS FLAVEOLA

Alien
6 inches (15 cm)

This attractive native of South America was introduced to Hawai'i in the 1960s. A pleasant song and brilliant yellow plumage make it a popular cage bird in many parts of the world. The Saffron Finch is a common bird on the leeward coast of the Big Island. It can also be seen on O'ahu and has recently been reported from both Maui and Kaua'i. It prefers dry areas, where it is usually seen foraging on the ground or in roadside weeds for seeds. A gregarious species, it is often seen in pairs or small groups. Like the Chestnut Munia, it occasionally forms large flocks. Adult males are bright yellow with yellow green backs and an orange or saffron wash on the forehead and crown. The bill is two-toned, with the lower mandible lighter in color. Females and immature birds are a dull yellowish gray.

The Big Island Country Club is an excellent place to look for small, introduced finches. They often feed on grasses that grow along the entrance road. Be careful not to obstruct the coming and going of golf course patrons. (JIM DENNY)

Saffron Finch, Sicalis flaveola. (JIM DENNY)

Yellow-fronted Canary

SERINUS MOZAMBICUS

Alien
4.5 inches (11 cm)

The Yellow-fronted Canary was imported from Africa in the 1960s. It has become established only on O'ahu and the Big Island. Most sightings are in the leeward areas. Adult males are greenish gray above and a bright yellow below. The crown and back of the head are gray. Yellow and black markings on the face form a large dark **X** when viewed from the front. The Yellow-fronted Canary travels in small flocks. It feeds on insects and seeds both on the ground and in trees. On O'ahu, the grasses in and around the Honolulu Zoo are good places to look for this species.

Yellow-fronted Canary, Serinus mozambicus. (JIM DENNY)

Yellow-faced Grassquit

TIARIS OLIVACEA

Alien
4 inches (10 cm)

This small finch is olive green above and grayish brown below. The breast and crown are black. The throat is yellow. Similar to the Yellow-fronted Canary, the black markings on the face form a large yellow **X** when viewed from the front. Females are somewhat duller in color and lack the bold face markings of the male. This introduced Central American native is currently established only on the island of O'ahu. Birds have been reported feeding in grasses along the roadsides above Wai'anae and in upper Pearl City.

Yellow-faced Grassquit, Tiaris olivacea. (MICHAEL WALTHER/O'AHU NATURE TOURS)

Red Avadavat

AMANDAVA AMANDAVA

Alien
4 inches (10 cm)

The Strawberry Finch, as some call it, often travels in small flocks. The call is a high-pitched "seet." It feeds on grass seeds, occasionally mixing with other seed-eating species like waxbills and mannikins. Males are red with white spots. Females and juveniles are a pale yellow brown. The bill and legs in both sexes are red. The Red Avadavat can be found in open areas where suitable grasses exist, often near water sources. This species was introduced from Southeast Asia in the early 1900s. It is now common but somewhat locally distributed on O'ahu (Kahuku aquaculture farms), the Big Island (Big Island Country Club and Pepe'ekeo), and Kaua'i (Po'ipū and Wailua). It has also been reported from the central valley on Maui.

Red Avadavat, Amandava amandava. (JIM DENNY)

Common Waxbill

ESTRILDA ASTRILD

Alien
4 inches (10 cm)

At only 4 inches, this attractive African native is one of the smallest birds in Hawai'i. It was first reported on O'ahu in the late 1970s, where it is now abundant and widespread.. The species is expanding to other islands as well. An increase in sightings has been reported from the Kailua-Kona area of the Big Island, from Po'ipū on Kaua'i, and from the central Maui plain. Any flocks of Chestnut Munias or Nutmeg Mannikins should be scanned for the presence of this species. Like many estrildid finches, Common Waxbills feed on grass seeds either on the ground or in tall weeds. The bird has a red bill and red facial mask that extends through the eye. The tail is black in adults, dull brown in juveniles.

Common Waxbill, Estrilda astrild. (JIM DENNY)

Black-rumped Waxbill

ESTRILDA TROGLODYTES

Alien
4 inches (10 cm)

This seldom-seen African native is known only from the North Kona District of the Big Island. It is so similar in size and appearance to the Common Waxbill that the two species are often confused. As the name suggests, the bird has a black rump. Other differences between these two red-browed waxbills are the under-tail coverts. They are black in the Common Waxbill and white in the Black-rumped Waxbill. The Black-rumped Waxbill also has less-prominent striations and a white lining to the outer tail feathers. Birds have been reported in the Pu'u Anahulu area and from the Big Island Country Club.

Black-rumped Waxbill, Estrilda troglodytes. (JACK JEFFREY)

Lavender Waxbill

ESTRILDA CAERULESCENS

Alien
4.5 inches (11 cm)

The Lavender Waxbill, or Lavender Firefinch as it is sometimes called, is a native of central Africa. This attractive seed-eating finch prefers dry habitats near water sources. Reports are increasing on the leeward coast of the Big Island. Good places to look for it are on the lawns of Kona hotels and in the grasses behind Keauhou Small Boat Harbor. In addition to the Big Island, a small population exists in the Hawai'i Kai area of O'ahu. Reports have also come from Maui near the Keālia Pond National Wildlife Refuge. The Lavender Waxbill is gray with a dark lavender bill. The rump and tail are a deep crimson red. It travels in small flocks that are difficult to see when feeding in the deep grass of a well-watered lawn.

Lavender Waxbill, Estrilda caerulescens. (JIM DENNY)

Orange-cheeked Waxbill

ESTRILDA MELPODA

Alien
4.5 inches (11 cm)

The Orange-cheeked Waxbill can only be seen on O'ahu and Maui. This African native was first reported near Diamond Head in the 1960s. The population has declined on O'ahu. Only a few sporadic sightings come from the windward side, particularly the Kāne'ohe area. On Maui, most reports come from the central valley. Look for them in the wild grasses at the Keālia Pond National Wildlife Refuge. Like the African Silverbill, the Orange-cheeked Waxbill constantly swishes its tail from side to side while feeding. This estrildid finch is light brown above and dusty white below. It has a red rump and bill and a conspicuous orange cheek patch.

Orange-cheeked Waxbill, Estrilda melpoda. (JIM DENNY)

Red-cheeked Cordonbleu

URAEGINTHUS BENGALUS

Alien
5 inches (13 cm)

This African native is one of several species of finches that were intentionally released from the Pu'u Wa'awa'a Ranch on the Big Island in the 1960s. The adults of both sexes are brown on the back and belly. The breast, face, and tail are blue. The adult male has an attractive red patch on the cheek. The Red-cheeked Cordonbleu is difficult to find. Most sightings are of birds associating with other seed-eating finches in the Pu'u Anahulu area. A small population was also present on O'ahu in the vicinity of Kapi'olani Park in the 1970s, but no reports have come from there in recent years.

Red-cheeked Cordonbleu, Uraeginthus bengalus. (GERHARD HOFMANN)

Java Sparrow

PADDA ORYZIVORA

Alien
6 inches (15 cm)

The Java Sparrow, with bold white cheeks and a thick pinkish bill, resembles a small puffin. Adult birds of both sexes are similar. Immature birds are duller overall, and the white facial patch is less obvious. It can often be seen in large flocks feeding on grass seeds, usually on the ground. The call is a loud, rapid "tik, tik, tik." This species will frequently dominate a bird feeder, with twenty or thirty birds at once not uncommon. This species is becoming increasingly abundant on all islands, especially O'ahu, Kaua'i, and the Big Island. The Java Sparrow is native to Indonesia.

Java Sparrow, Padda oryzivora. (JIM DENNY)

African Silverbill

LONCHURA CANTANS

Alien
4.5 inches (11 cm)

This small estrildid finch was once known as the Warbling Silverbill. It is not known how the species arrived in the Islands, but it was first recorded on the Big Island in the 1970s. It is most frequently seen in dry areas, where it is attracted to water sources. The call, which it utters in flight, has been described as a metallic "tic-tic-tic" sound like that of two coins being clicked together. It occasionally feeds in mixed flocks with Chestnut Munias and Nutmeg Mannikins. The black tail, which moves constantly from side to side, contrasts strongly with the pale underparts and brown upper parts. The thick conical bill is light blue. Although the African Silverbill is present on all the main islands, it is more numerous on the Big Island, Moloka'i, and Lāna'i.

African Silverbill, Lonchura cantans. (JIM DENNY)

Chestnut Munia

LONCHURA ATRICAPILLA

Alien
4.5 inches (11 cm)

The Chestnut Munia, a native of Southeast Asia, is an escaped cage bird that was first recorded on O'ahu in 1941. Also known as the Chestnut Mannikin or Black-headed Munia, this species is now present on all the Islands except Moloka'i and the Big Island. On O'ahu and Maui, look for it in the tall green grass along roadsides and agricultural ditches. On Kaua'i, it is so plentiful that it may well be the island's most numerous lowland bird, with flocks so large they resemble a cloud of bees. A Chestnut Munia will land on the tall seed-bearing stalk of a weed to bend it over with its weight. It then walks out to the end to eat the seed. As the name describes, the Chestnut Munia is chestnut brown with a black head. The large conical bill is silver with a hint of blue.

Chestnut Munia, Lonchura atricapilla. (JIM DENNY)

Nutmeg Mannikin

LONCHURA PUNCTULATA

Alien
4.5 inches (11 cm)

The Nutmeg Mannikin, also known as the Ricebird or Scaly-breasted Munia, was introduced from Southeast Asia in 1865. It is now present on all the Islands. This small bird was a serious agricultural pest when rice farming was prevalent in Hawai'i. Flocks containing hundreds of birds would invade a field and destroy much of the crop. Nutmeg Mannikins feed by walking up a grass stem to get at the seed clusters on the ends. Like the Chestnut Munia, Nutmeg Mannikins feed in large flocks, but they are more wary and difficult to approach. The Nutmeg Mannikin is seen at higher altitudes than the Chestnut Munia and even inhabits open grassy areas in the high wet native forest. Adults have a brown head and heavy conical black bill. The underneath is pale brown and heavily scalloped.

Nutmeg Mannikin, Lonchura punctulata. (JACK JEFFREY)

The majestic volcano Mauna Loa serves as a backdrop to the extensive grasslands of the Big Island's upper slopes. A drive along the roads here will often produce several species of introduced game birds and raptors. (JIM DENNY)

COUNTRY BIRDS

MANY OF THE BIRDS SEEN in urban and forest habitats in Hawai'i also thrive in the countryside, but the open spaces are especially favored by several species of game birds and raptors. Game birds were first released in the islands by visiting ships that hoped to ensure a good supply of food for subsequent stopovers in Hawai'i. In the 1920s more game birds were introduced. Additional species came in the 1950s and 1960s as the government attempted to establish birds for recreational hunting. Most adapted well to the lava flows, pasturelands, alpine slopes, and lowland scrub of Hawai'i. Not all attempts were successful, however. The Northern Bobwhite, Sharp-tailed Grouse, and the Montezuma Quail are among several that failed to sustain breeding populations. Illustrated in this guide are eleven species of game birds that have become well established.

Also illustrated in this group are a native hawk, a native owl, and one introduced owl. Three migrant raptors, the Peregrine Falcon, Osprey, and Northern Harrier, are described as well. Records exist of several other raptors in the state, but such sightings are extremely rare. They include the Merlin, Rough-legged Hawk, and Golden Eagle.

Completing the category of country birds are two lark species often associated with farmlands and meadows.

Hawaiian Hawk (*'Io*)

BUTEO SOLITARIUS

ENDANGERED
Endemic
18 inches (46 cm)

There is fossil evidence that the Hawaiian Hawk once lived on Kaua'i, Moloka'i, and O'ahu, but today this endemic raptor is known to reside only on the Big Island. The *'Io*, the Hawaiian name for the bird, can be seen islandwide soaring on the thermals high above pastures, lava flows, and forests. Like the native owl, this native hawk feeds mostly on insects, rodents, and birds. It generally chooses the same nest site year after year. The Hawaiian Hawk has two color morphs—a light and a dark phase. The feet and legs are yellow in adults and greenish yellow in juvenile birds.

Hawaiian Hawk ('Io), Buteo solitarius *(light morph).* (JACK JEFFREY)

Hawaiian Hawk ('Io), Buteo solitarius *(dark morph).* (JACK JEFFREY)

Northern Harrier

CIRCUS CYANEUS

Migrant
23 inches (58 cm)

Also known as the Marsh Harrier, this large North American hawk is a rare visitor to Hawai'i. In recent years, birds have been reported from Maui, Moloka'i, and O'ahu. This species readily crosses open water, a characteristic that may explain its occasional appearance in the Islands. It feeds on rodents and behaves much like the native *Pueo* when searching for prey. It cruises low over grassy fields and marshes and hovers in place when trying to locate the sound of mice on the ground. This raptor has a prominent white rump, a feature that readily separates it from the Hawaiian Hawk and *Pueo.*

Northern Harrier, Circus cyaneus. (PETER LATOURETTE)

Osprey

PANDION HALIAETUS

Migrant
23–25 inches
(56-64 cm)

The Osprey can be found on nearly every continent. Sightings in Hawai'i are rare, but the bird has been reported from all the main islands. This large raptor is also known as the Fish Hawk or Fish Eagle. Unlike most raptors, the wings are bent at the wrist. With a wingspan of nearly 6 feet (2 m), a high-flying Osprey resembles the silhouette of a large seabird. The Osprey feeds entirely on fish. It hunts by hovering over its victim and then plunging claws first into the water to seize it. In Hawai'i it has been observed hunting over both inland and coastal waters. Except for the white head, the Osprey is dark above. Below, the body and coverts are nearly all white. The primaries and secondaries are darkly barred. There is a dark patch at the bend of the wrist. The eyes of the adult are yellow. A broad dark eye stripe extends from the base of the bill to the back of the crown.

Osprey, Pandion haliaetus. (JIM DENNY)

Peregrine Falcon

FALCO PEREGRINUS

Migrant
16 inches (41 cm)

A highly migratory species, this large falcon is found worldwide. Nearly every year sightings are reported in the Hawaiian Islands. Birds have been recorded over forests, wetlands, seabird colonies, and even in urban areas on all the Islands except Lāna'i. An efficient predator, the Peregrine Falcon is known for the incredible speed with which it pursues birds on the wing. Although plumage can vary, all birds have a wide sideburn on the face that extends well below the eye. The wings are long and taper to a point.

Peregrine Falcon, Falco peregrinus. (JIM DENNY)

Hawaiian Owl *(Pueo)*

ASIO FLAMMEUS SANDWICHENSIS

Endemic
13–17 inches
(33–43 cm)

The ancient Hawaiians regarded the Hawaiian Short-eared Owl, or *Pueo*, with reverence. Fossil bones of this species do not appear in Hawaiian deposits until after Polynesian arrival. Scientists theorize that earlier attempts at colonization may have failed because the open grassland habitat favored by this species was probably not present until after humans altered the Hawaiian landscape. The *Pueo* hunts around daybreak, at sunset, and sometimes at midday. Although it frequents fields and pastures, this endemic species can also be found in the native forest. Its diet is mostly rodents and insects, but the *Pueo* does prey upon both native and alien birds. When hunting it hovers above its prey before diving down to attack. Deep, penetrating yellow eyes and a round facial disk distinguish this species from the Barn Owl. The legs and feet are covered in feathers.

Hawaiian Owl (Pueo), Asio flammeus sandwichensis. (JACK JEFFREY)

Barn Owl

TYTO ALBA

Alien
16 inches (41 cm)

The Barn Owl can be found throughout Hawai'i. Sugar planters introduced it in the late 1950s in an effort to control rats in the fields. Rodents do make up a large part of its diet; but the Barn Owl also preys on birds and has been responsible for nesting failures of seabirds at Kīlauea Point National Wildlife Refuge. Although it can occasionally be seen hunting during daylight hours, this species is mostly a night predator. The harsh raspy call of the Barn Owl is a common sound of the night in rural areas of Hawai'i. Sexes differ in size and in the amount of white in the face, but both male and female of this species have dark eyes in a heart-shaped facial disk. The only other owl present in Hawai'i is the native *Pueo*, which has yellow eyes in a round dark face.

Barn Owl, Tyto alba. (JIM DENNY)

Cattle Egret

BUBULCUS IBIS

Alien
20 inches (51 cm)

State officials imported the Cattle Egret from Florida in the late 1950s in an attempt to control insect pests on island cattle ranches. It is now firmly established on all the Islands. Some residents call it the "rubbish dump bird" because it congregates there by the hundreds. It frequently follows mowing machinery in parks and fields to ingest the insects that are stirred up. Cattle Egrets are known to eat almost anything and pose a threat to the nestlings of waterbirds and seabirds. Cattle Egrets rarely wade in the water but often choose to roost near it in large assemblies of a hundred birds or more. As they fly to their communal roosting sites late in the day, flocks of these white birds form beautiful scenes against the lush greenery of Hawai'i's mountains.

Cattle Egret, Bubulcus ibis. (JACK JEFFREY)

Black Francolin

FRANCOLINUS FRANCOLINUS

Alien
14 inches (35 cm)

This handsome South Asian native was introduced to Hawai'i as a game bird in 1959. It is black overall but boldly marked with many white spots. The short fanlike tail is finely streaked with alternating black and white lines. A solid wide chestnut band encircles the neck. A conspicuous patch of white on the cheek extends from below the eye to the back of the neck. The Black Francolin prefers the same dry habitat as the Erckel's Francolin, but it is shyer and thus more difficult to observe. The loud, buzzy, high-pitched call is given mostly in the early morning and late afternoon. The call often ends with the phrase "bee-vee-dee," causing some local hunters to refer to it as the BVD bird. The Black Francolin is common in suitable habitat on all islands except Lāna'i.

Black Francolin, Francolinus francolinus. (JIM DENNY)

Gray Francolin

FRANCOLINUS PONDICERIANUS

Alien
13 inches (33 cm)

A native of India, this small short-tailed game bird was introduced to all the Hawaiian Islands in 1958. It is now common in lowland areas on Lāna'i, Moloka'i, Maui, and the Big Island. On O'ahu, a few birds reside on the slopes of Diamond Head and in the vicinity of Pearl Harbor. On Kaua'i, the species is rarely seen. Above, the bird is covered almost entirely with light brown rectangles bordered with creamy white lines. Below, it is finely barred with thin alternating brown and white lines. A dull yellow throat is sometimes visible. Like the other two francolin species in Hawai'i, the Gray Francolin likes to venture out onto the roads in early morning and late afternoon. It is also fond of foraging near and under *kiawe* trees. In rural areas where the bird is abundant, the "ki-oh, ki-oh, ki-oh" call of this species is a frequent component of the morning cacophony of birdsongs.

Gray Francolin, Francolinus pondicerianus. (JIM DENNY)

Erckel's Francolin

FRANCOLINUS ERCKELII

Alien
16 inches (41 cm)

This large, short-tailed game bird, a native of northern Africa, was introduced to Hawai'i in 1957. It is has been reported on all of the main islands, but most sightings occur on Kaua'i and the Big Island. The bird is streaked with chestnut and gray. A white throat is sometimes visible. The crown is solid chestnut lined with a black superciliary stripe. It favors high grasses in dry upland habitats but can be found in the wet forest as well. The Erckel's Francolin likes to venture out onto the roads in early morning and late afternoon. Some locals call this species the Laughing Bird. When it calls from concealment, it gives the impression that someone is having a good guffaw at your expense.

Erckel's Francolin, Francolinus erckelii. (JACK JEFFREY)

Chukar

ALECTORIS CHUKAR

Alien
14 inches (36 cm)

This species was introduced from Asia in 1923. Males and females are similar in plumage. In general, it looks like a large short-tailed quail. It has a broad black stripe extending from the forehead through the eye and down onto the cream-colored throat. The sides of the body are boldly marked with black stripes. Both bill and feet are red. The eye is black with a red eye ring. Like quails, the Chukar prefers dry grassy shrubland, particularly near water sources. It has adapted well to the arid conditions of Hawai'i's high volcanoes. On Maui, it can often be seen perched on rocks alongside the road near the summit of Haleakalā. On the Big Island, look for them along the Saddle Road. The Chukar is common on all islands except Kaua'i and O'ahu.

Chukar, Alectoris chukar. (JACK JEFFREY)

California Quail

CALLIPEPLA CALIFORNICA

Alien
10 inches (25 cm)

This plump, short-tailed game bird was introduced to Hawai'i before 1860. As the name suggests, it is native to the western United States. Adult males have a black throat and brown crown, each bordered by a white stripe. The belly is scaled with a diffuse chestnut patch. The most prominent feature, however, is the black teardrop plume found on the head of both male and female birds. Females are duller overall. The California Quail prefers dry grassy shrubland. It is rare on all the Islands expect the Big Island and Moloka'i. An excellent roadside location to see this bird on that island is the Mauna Kea State Park along the Saddle Road. In the early morning and late afternoon, it gathers on the road in large coveys. Listen for the distinct three-note call, "chi-ca-go," the second syllable higher in pitch.

Mauna Kea State Park, on the Big Island's Saddle Road, is a good place to see the introduced California Quail. Early and late in the day, birds often venture out from the surrounding scrub to gather on the lawn in large coveys. (JIM DENNY)

California Quail, Callipepla californica. (JIM DENNY)

Gambel's Quail

CALLIPEPLA GAMBELII

Alien
10 inches (25 cm)

Native to the southwestern United States, the Gambel's Quail was first introduced to Hawai'i in the 1920s as a game bird. In appearance, it is very similar to the California Quail and, in fact, may hybridize with it. The best way to separate adult males of the two species is to pay attention to the belly. The Gambel's Quail has a large dark spot on cream-colored nonscaled underparts. This species also prefers grassy shrubland but can tolerate drier habitats than the California Quail. Most reports are from Lāna'i, where it is considered common. Listen for a four-note call, higher in pitch than the call of its California cousin.

Hawai'i's smallest game bird, the Japanese Quail *(Coturnix japonica)*, has also been introduced to all of the main islands. This small, pale brown bird prefers large open spaces that are covered in short grass. It is difficult to flush and blends into its habitat so effectively that it is rarely seen. Most reports, based on its wheezy "chik-whir" call, come from pastures in the Princeville area of Kaua'i, the slopes of Mauna Kea, and the northwestern slope of Haleakalā.

Gambel's Quail, Callipepla gambelii. (JIM DENNY)

Wild Turkey

MELEAGRIS GALLOPAVO

Alien
46 inches (117 cm)

The Wild Turkey is among the earliest birds introduced to Hawai'i. It arrived from North America nearly 200 years ago. It is one of the largest game birds in the state. The body of the adult male is an iridescent mix of dark brown, black, and bronze. The head is pink and red with hanging wattles. The legs are spurred. More sightings come from the Big Island than any other location. The Saddle Road and the Big Island Country Club are reliable places to observe this species. Numbers are increasing in the upper Māhā'ulepū area of Kaua'i.

Wild Turkey, Meleagris gallopavo. (JACK JEFFREY)

Common Peafowl

PAVO CRISTATUS

Alien
100 inches (254 cm)

More commonly known as a Peacock, this beautiful bird was brought from India in 1860. Adult male birds are an iridescent blue and green with white facial markings. A prominent crest adorns the head. When the tail is held erect, the Common Peafowl is one of the most stunning birds in nature. There are a few wild populations dispersed in rural areas of the Hawaiian Islands, but most are shy and not readily observable. Habituated birds have been reported, however, in the Makawao area of Maui, in the O'ahu neighborhoods of Mililani Mauka and Kahalu'u, and roaming the grounds along the bank of the Wailua River near Smith's Tropical Gardens on Kaua'i.

Common Peafowl, Pavo cristatus. (JIM DENNY)

Ring-necked Pheasant

PHASIANUS COLCHICUS

Alien
33 inches (84 cm)

A favorite of local hunters, this large, long-tailed bird was introduced to the Islands from Asia over a hundred years ago. The plumage of the adult male is striking. Overall, it is an iridescent chestnut bronze spotted with black, white, and brown. The long, pointed tail is barred. The face is chickenlike with a large red patch of skin around the eye. The head and neck of the Ring-necked Pheasant vary between glossy dark green and purple. A prominent bright white ring encircles the lower neck. The call is a loud two-note "kuk-kock" that can be heard from long distances. This adaptable species can be seen in many rural habitats both dry and wet. A search of roadside pastures on any island is likely to produce this beautiful bird. Another race, the Green Pheasant *(Phasianus colchicus versicolor)*, is also present in Hawai'i but rarely seen. The male Green Pheasant is green brown above and an iridescent blue green below. The white collar is absent. Some sources consider it a separate species.

Ring-necked Pheasant, Phasianus colchicus. (JACK JEFFREY)

Chestnut-bellied Sandgrouse

PTEROCLES EXUSTUS

Alien
12 inches (30 cm)

The seldom-seen Chestnut-bellied Sandgrouse is native to the arid regions of Africa, India, and Asia. It was introduced to Hawai'i as a game bird in 1961. It is similar in size and shape to the Rock Pigeon. The attractive male is sandy colored with a long, thin, pointed tail and a black band that separates the lower breast from the belly. It is established only on the Big Island. Look for it feeding on seeds along the side of the road or in pastures in the South Kohala District. Birds are also occasionally seen in small flocks flying rapidly to and from water sources near the western end of the Saddle Road.

Chestnut-bellied Sandgrouse, Pterocles exustus.
(MICHAEL WALTHER/O'AHU NATURE TOURS)

Sky Lark

ALAUDA ARVENSIS

Alien
7.5 inches (19 cm)

The Eurasian Sky Lark was introduced to Hawai'i from England in the 1860s. The breast and belly are white, and the back is streaked with brown and white. In this respect, it resembles a large House Finch. Unlike the House Finch, however, the bill is longer and more slender. In flight, the outer tail feathers are conspicuously white. The song of the Sky Lark is a lengthy complicated series of trills and warbles. It often sings in flight. Sky Larks have been reported from all the Islands, but they are most common on the high slopes of Maui and the Big Island. Look for them along the roadside as they take flight to avoid approaching automobiles.

Sky Lark, Alauda arvensis. (JACK JEFFREY)

Western Meadowlark

STURNELLA NEGLECTA

Alien
9.5 inches (24 cm)

As the name suggests, the Western Meadowlark is a bird of open spaces and pasturelands. Native to western North America, this gifted songster was introduced to O'ahu and Kaua'i in 1931, but only the Kaua'i population has survived. To date, Kaua'i remains the only place in the state where the Western Meadowlark can be seen. The species has a yellow breast accented with a prominent black **V**. In flight, white feathers are obvious on either side of the tail. The meadowlark likes to sing from high perches, so look for it atop fence posts in pastures from Mānā to Hanalei. Excellent places to see this bird are along the north end of the Līhu'e airport near the NOAA weather station and at the Burns Field Airstrip in Hanapēpē.

Western Meadowlark, Sturnella neglecta. (JACK JEFFREY)

Between the two scenic viewpoints of Kalalau Valley, Kaua'i's Kōke'e Road travels through almost half a mile (0.8 km) of thick native rain forest. The trees and shrubs along the roadside are excellent habitat for four Kaua'i endemics: 'Apapane, 'Anianiau, 'Amakihi, *and* 'Elepaio. (JIM DENNY)

FOREST BIRDS IN HAWAI'I ARE a rich blend of native and introduced species. To see native forest birds, one must go *mauka,* or up the mountains. Disease-bearing mosquitoes and the lack of ample native forest at lower elevations has restricted most endemic forest birds to the mountains. Hosmer Grove on Maui, Kīpuka 21 on the Big Island, and the Kalalau scenic viewpoints on Kaua'i are easily accessed locations where one can still see the native *'Apapane* and *'Amakihi* without much difficulty. Off-road hikes into the native forest will surely produce others. Several specific trails are listed in this section. Twenty-two species or subspecies of native forest birds are illustrated.

Introduced birds are a small part of this group, but their impact is substantial. In some island forests, they far outnumber native species. Some of the ships that visited the Islands in the 1800s brought with them forest birds from their travels. Then, beginning in the

1930s, an avian society named the Hui Manu actively imported birds from around the globe in an attempt to replace the dwindling populations of native forest birds. The consequence of over 200 years of introductions has contributed greatly to the number of alien bird species in our forests. In addition to the eight introduced forest birds illustrated in this section, there were many other releases that did not take well to the Hawaiian Islands. Among them were the Indigo Bunting from North America, the Ryukyu Robin from the subtropical Asian islands, and the Black-throated Laughingthrush from China.

Today, the Hawai'i State Department of Agriculture has regulations regarding the importation of alien species. Even if a permit is obtained to bring in a bird on the "conditionally approved" list, that imported bird must remain in captivity. These laws are necessary to protect endemic birds and other native species. There are those who would love to make Hawai'i more of a paradise by having a colorful parrot in every tree, but it is important to remember that each new released or escaped bird carries with it the potential of bringing a new disease or parasite to afflict dwindling native species or new competitors for food and space in the native forest.

Hawaiian Crow (*'Alalā*)

CORVUS HAWAIIENSIS

ENDANGERED
Endemic
20 inches (51 cm)

The Hawaiian Crow, or *'Alalā,* is a large, all-black bird with a heavy black bill. The species is found only on the Big Island, but no longer exists in the wild. Predation, disease, habitat loss, and other factors have reduced the number of *'Alalā* to about fifty birds. All of these are in captive breeding facilities on Maui and the Big Island. Released birds have not fared well, and it is doubtful that the *'Alalā* will ever again thrive as a wild population. Like most crows, it is a noisy social species that moved through the forest in groups. It has a varied diet that includes fruit, seeds, insects, mice, and even carrion.

Hawaiian Crow ('Alalā), Corvus hawaiiensis. (JACK JEFFREY)

Hawai'i *'Elepaio*

CHASIEMPIS SANDWICHENSIS SANDWICHENSIS

Endemic
5.5 inches (14 cm)

Hawai'i 'Elepaio, Chasiempis sandwichensis sandwichensis *(Mauna Kea).* (JACK JEFFREY)

The Big Island race of *'Elepaio* is the most boldly marked of the three subspecies in Hawai'i. Like the other island forms, adults have a prominent white rump and two white wing bars. The Hawai'i *'Elepaio* has additional white speckling on the breast, throat, and head. In females, the throat is nearly all white. In males, it is white but speckled with black. Some birds in the drier forest on Mauna Kea are paler overall with much more white on the head. As with other endemic forest birds, the Hawai'i *'Elepaio* is mostly a bird of the high-altitude forests. However, a few sightings have occurred at lower elevations. Kīpuka 21, Pu'u 'Ō'ō Trail, and Manukā State Wayside Park are good places to search for this perky flycatcher. The classic tail-up pose is usually the first key to identifying the bird.

Hawai'i 'Elepaio, Chasiempis sandwichensis sandwichensis *(volcano).* (JACK JEFFREY)

O'ahu *'Elepaio*

CHASIEMPIS SANDWICHENSIS IBIDIS

ENDANGERED
Endemic
5.5 inches (14 cm)

The O'ahu race is similar to the Big Island volcano variety. It also has prominent white wing bars and a splotchy white throat. However, the O'ahu *'Elepaio* has more white underneath and less brown streaks on the breast. The population of the O'ahu *'Elepaio* is in decline. It now inhabits only the upper Wai'anae and Ko'olau mountains. There is currently no easily reached location where this endangered flycatcher can be seen. A hike along the precipitous 'Aiea Ridge Trail or into Kuli'ou'ou Valley might possibly produce the bird.

O'ahu 'Elepaio, Chasiempis sandwichensis ibidis. (JACK JEFFREY)

Kaua'i *'Elepaio*

CHASIEMPIS SANDWICHENSIS SCLATERI

Endemic
5.5 inches (14 cm)

Kaua'i 'Elepaio, Chasiempis sandwichensis sclateri *(juvenile).* (JIM DENNY)

The *'Elepaio* is a favorite of many birders in Hawai'i because it shows little fear of humans. An imitation of its call will often entice juveniles to approach close enough to touch. The Kaua'i subspecies is gray brown above and pale below with a diffuse orange wash on the breast. Immature birds are rusty colored. In its search for insects, this adaptable bird feeds in a variety of habitats. It can dart through the understory to catch flies and moths in the air or forage high in the canopy for arthropods hidden among the leaves. As is common with many Old World flycatchers, the nest of the *'Elepaio* is spotted with lichens. A walk along the road between the two Kalalau scenic viewpoints or along the Pihea Trail will often produce this friendly bird.

Kaua'i 'Elepaio, Chasiempis sandwichensis sclateri *(adult).* (JIM DENNY)

ʻŌmaʻo

MYADESTES OBSCURUS

Endemic
7 inches (18 cm)

This large endemic thrush is found only on the Big Island, where it is common in the native wet forest on the eastern slopes of Mauna Kea and Mauna Loa. A few birds have also been reported in drier areas on these two mountains. The other main islands in Hawaiʻi each had a distinct species of endemic thrush. All except the *Puaiohi* and *ʻŌmaʻo* are extinct. Light brown above and gray below, the *ʻŌmaʻo* is difficult to see when perched motionless on a large *ʻōhiʻa* or *koa* tree. Fortunately, it has a loud fluidlike song and a raspy call that often gives away its position. The *ʻŌmaʻo* also has the curious habit of quivering its wings when perched. Look for it feeding on *ʻōlapa* or *pilo* fruit in Hawaiʻi Volcanoes National Park or at Kīpuka 21 on the Saddle Road.

ʻŌmaʻo, Myadestes obscurus. (JACK JEFFREY)

Small Kaua'i Thrush *(Puaiohi)*

MYADESTES PALMERI

ENDANGERED
Endemic
7 inches (18 cm)

The *Puaiohi*, or Small Kaua'i Thrush, is a rare endemic frugivore found only on the island of Kaua'i. It is dark gray brown above and pale gray brown below. The legs are pink; the bill is slender and black. A faint, white eye ring is present. Immature birds are heavily scalloped below. The *Puaiohi* prefers deep, narrow valleys in the wet forest, where it nests in cavities in the steep, fern-covered valley walls. The likelihood of seeing this species in the Kōke'e area is extremely poor; however birds propagated in captivity are occasionally released into the Kawaikōī Stream valley. The chance of an encounter increases the farther one hikes into the Alaka'i Swamp. The upper end of the rugged Mōhihi-Wai'alae Trail near the Koai'e Stream crossing has been a good area in recent years to search for this illusive species.

Small Kaua'i Thrush (Puaiohi), Myadestes palmeri. (JACK JEFFREY)

Palila

LOXIOIDES BAILLEUI

ENDANGERED
Endemic
7.5 inches (19 cm)

The *Palila* has a heavy finchlike bill and a dark mask. The head and breast of the adult male are bright yellow. It is white underneath with a gray back. The *Palila* feeds almost exclusively on immature seedpods of the *māmane* tree and on the fruit and flowers of *naio.* When feeding on *māmane*, the bird pulls a pod from the tree, flies to a nearby branch, and, while gripping the pod with its feet, it rips open the tough hull to get at the seeds inside. The process often takes several minutes to accomplish. The call, a two- or three-note whistle, is extremely useful in locating this precious native honeycreeper. The endangered *Palila* inhabits an area of *māmane* and *naio* forest high on the dry slopes of Mauna Kea known as Puʻu Laʻau. Access to Puʻu Laʻau is not restricted, but a 4x4 vehicle is required to enter the area.

Palila, Loxioides bailleui. (JACK JEFFREY)

"Palila Road" is located near the 43-mile marker on the Big Island's Saddle Road. This rugged four-wheel drive dirt road travels 4 miles (6.4 km) upslope to the māmane *and* naio *forest at Puʻu Laʻau. The area is home to the* Palila*, an endangered native honeycreeper.* (JIM DENNY)

'Akakane

LOXOPS COCCINEUS COCCINEUS

ENDANGERED
Endemic
4 inches (10 cm)

'Akakane *(Hawai'i* 'Ākepa*)*, Loxops coccineus coccineus *(female)*. (JACK JEFFREY)

The male *'Akakane,* better known in the Hawai'i birding community as the Hawai'i *'Ākepa,* is bright orange with a long tail, notched at the end. Like the *'Akeke'e,* the bill is crossed or off-set at the tip. The bill is a pale yellow orange. The legs and feet are black. The Hawai'i *'Ākepa* often forages in mixed feeding flocks with the Hawai'i Creeper and *'Akiapōlā'au.* The Hawai'i *'Ākepa* feeds on insects found in leaf clusters high in the *'ōhi'a* canopy or on larvae found in infest-ed fruit like the *'ōhelo,* a common native shrub that grows in the subcanopy. The most reliable place to find the *'Akakane* is the Hakalau Forest National Wildlife Refuge. Access is prohibited without permission.

'Akakane (*Hawai'i* 'Ākepa), Loxops coccineus coccineus. (JACK JEFFREY)

ʻAkekeʻe

LOXOPS CAERULEIROSTRIS

ENDANGERED
Endemic
4.5 inches (11 cm)

This small green bird is referred to by some as the Kauaʻi *ʻĀkepa*. It resides only at high elevations on that island. A short, bluish gray conical bill and black lores set it apart from the similar-sized *ʻAnianiau*. The long tail is notched at the end. It feeds high in the canopy, where it forages through the buds and terminal leaf clusters of *ʻōhiʻa* trees in search of insects. Its bill, crossed at the tip, is uniquely adapted to open the closely spaced leaves. It rarely takes nectar. The *ʻAkekeʻe* often travels in small family groups that call back and forth to each other as they feed. It is a busy bird, constantly moving in and out of the leaves. The Hawaiian word *ʻākepa* literally means active. This species certainly fits that description. Until recently the *ʻAkekeʻe* could occasionally be seen foraging in the *ʻōhiʻa* trees at the Kalalau scenic viewpoints. A sighting there is still possible, but the population has mysteriously declined to the point where the Kauaʻi *ʻĀkepa* is now a very difficult bird to find.

ʻAkekeʻe, Loxops caeruleirostris. (JIM DENNY)

'Anianiau

MAGUMMA PARVUS

Endemic
4 inches (10 cm)

The *'Anianiau* is the smallest of all Hawaiian honeycreepers. Males are bright yellow, and females are yellowish green. Its short, slightly curved, pink bill is adapted to feed on the flowers produced by such endemic plants as the *'ōhelo, kanawao,* and *naupaka kuahiwi. 'Anianiau* often feed in a circuit. Birds return to feed at the same plant and even the same flower. This small species can readily be identified by its call—a soft, slurred, two-note whistle rising in pitch. It can only be seen in the upland native forest on the island of Kaua'i. The *'Anianiau* is still one of the more common endemic birds encountered along the Pihea Trail, but the population is not as robust as a decade ago.

'Anianiau, Magumma parvus. (JIM DENNY)

Kaua'i *'Amakihi*

HEMIGNATHUS KAUAIENSIS

Endemic
4.5 inches (11 cm)

This small yellow green bird is one of Kaua'i's most common forest endemics. The *'Amakihi* in Hawai'i exist in several distinct species: one on O'ahu, one on Kaua'i, and a third that has been divided further into two subspecies, one representing the Big Island and the other the combined Maui, Moloka'i, and Lāna'i group of islands. The Kaua'i *'Amakihi* has the longest and thickest bill of the four. It can be recognized easily by its call—two separate notes, the second higher in pitch. With its multipurpose bill it feeds on a variety of sources. It seeks insects beneath the bark of trees, takes nectar from native flowers in the understory, and has even adapted to feeding on introduced fruits like java plum and alien flowers like that of the banana poka. A hike along the Pihea Trail often produces the bird.

Kaua'i 'Amakihi, Hemignathus kauaiensis. (JIM DENNY)

O'ahu *'Amakihi*

HEMIGNATHUS FLAVUS

Endemic
4.5 inches (11 cm)

The O'ahu *'Amakihi* is similar to the Big Island race but has a slightly heavier bill. Two white wing bars are present on females and juveniles. Some males have a yellow eyebrow and more yellow on the breast. The song is slower than that of the *'Amakihi* on other islands. A hike along any trail that ventures into native forest is likely to turn up this bird. The 'Aiea Ridge Trail has been reliable in recent years. The O'ahu *'Amakihi* is an adaptable bird and has developed some resistance to avian malaria. It can occasionally be seen even at low elevations feeding among alien vegetation. Look for it in Wa'ahila Ridge State Recreation Area in St. Louis Heights, on Round Top Drive, or at Lyon Arboretum.

O'ahu 'Amakihi, Hemignathus flavus. (GEORGE ARMISTEAD)

Hawai'i *'Amakihi*

HEMIGNATHUS VIRENS VIRENS

Endemic
4.5 inches (11 cm)

The Big Island *'Amakihi* differs somewhat from the Kaua'i form. The color is similar (although some observers think that the Hawai'i Island bird is brighter green), but the bill is shorter, thinner, and not as down-curved as *Hemignathus kauaiensis*. It is also similar in size to the introduced Japanese White-eye, a species that is abundant in the native forests of Hawai'i. To separate the two it is important to remember that none of Hawai'i's small endemic honeycreepers has a white eye ring. Like the O'ahu *'Amakihi*, the Big Island *'Amakihi* is somewhat resistant to avian malaria. A few birds have been seen at low elevations. Easy roadside locations to see the Hawai'i *'Amakihi* are Volcanoes National Park and Kāpuka 21.

Hawai'i 'Amakihi, Hemignathus virens virens. (JIM DENNY)

Maui *'Amakihi*

HEMIGNATHUS VIRENS WILSONI

Endemic
4.5 inches (11 cm)

The Maui Nui (an island group that encompasses Maui, Moloka'i, and Lāna'i) subspecies of *'Amakihi* is virtually indistinguishable from the Big Island form. The only noticeable difference is a single faint white wing bar seen only in immature Maui birds. Most literature simply refers to the two subspecies as the Hawai'i *'Amakihi*. This may be Hawai'i's most easily seen honeycreeper. Habituated birds routinely feed on the picnic tables at Hosmer Grove. With a little enticement, the Maui *'Amakihi* at this small campground will even eat from the hand.

Hosmer Grove is one of the easiest places in Hawai'i to see native honeycreepers. Located at 6,800 feet (2,072 m), just inside the Haleakalā National Park on the island of Maui, this small camping area has an abundance of flowering māmane, *a valuable nectar source for the curved-billed* 'I'iwi *and Maui* 'Amakihi. *The insectivorous Maui* 'Alauahio, *or Maui Creeper, is also common here. Hosmer Grove is the location of the trail head into the restricted Waikamoi Preserve.* (JIM DENNY)

Maui 'Amakihi, Hemignathus virens wilsoni. (JIM DENNY)

ʻAkiapōlāʻau

HEMIGNATHUS MUNROI

ENDANGERED
Endemic
5.5 inches (14 cm)

The endangered *ʻAkiapōlāʻau* can only be found in the high native forests of the Big Island. This beautiful short-tailed honeycreeper is a thrill to watch when it is foraging among the trees and shrubs. With behavior much like that of a woodpecker, it hammers away with the lower mandible at larva-infested branches to find the grub hidden within. It then uses the long, curved upper mandible to dig the larva out. Males are green above. The head and abdomen are yellow. Females are duller. The bill and the legs are black in both sexes. The lores are black. Most sightings of this uncommon bird are from Hakalau Forest National Wildlife Refuge (where access is restricted) or along the Puʻu ʻŌʻō trail. There have, however, been rare reports of this bird from Kīpuka 21 on the Saddle Road and from Puʻu Laʻau.

ʻAkiapōlāʻau, Hemignathus munroi. (JIM DENNY)

Kaua'i Creeper (*'Akikiki*)

OREOMYSTIS BAIRDI

ENDANGERED
Endemic
5 inches (13 cm)

The *'Akikiki* is found only on the island of Kaua'i. It feeds by creeping along the trunks and limbs of trees, where it probes the moss or crevices in the bark for hidden insects. The Kaua'i Creeper, as it is often called, is dark gray above and white below. The bill and legs are pink. The tail is short. It travels through the wet forest in pairs or small family groups, the individuals of which call to each other as they feed. It occasionally associates with the *'Akeke'e* and *'Anianiau* in mixed-species feeding flocks. The call is a very short "sweet" and resembles closely the call of the *'Akeke'e*. The *'Akikiki* is becoming increasingly rare. The chance of an encounter increases the farther one hikes into the Alaka'i Swamp. The upper end of the rugged Mōhihi-Wai'alae Trail near the Koai'e Stream crossing has been a good area in recent years to search for this precious honeycreeper.

Kaua'i Creeper ('Akikiki), Oreomystis bairdi, (JIM DENNY)

Hawai'i Creeper

OREOMYSTIS MANA

ENDANGERED
Endemic
4.5 inches (11 cm)

No Hawaiian name is known for this small insectivore that is found only in the high mountain forests of the Big Island. The bird is deep olive green above and slightly paler below. The legs and bill are slate gray. It has broad black lores. It is similar in appearance to the Hawai'i *'Amakihi,* but the bill is much straighter. Like its Kaua'i cousin, the *'Akikiki*, the Hawai'i Creeper moves up and down the trunks and branches of trees and large shrubs searching among the bark for spiders, beetles, and insect larvae. It frequently moves through the forest in the company of other small endemic honeycreepers like the *'Amakihi*, *'Ākepa*, and *'Akiapōlā'au.* The incessant begging "wheet-wheet-wheet" call of juveniles is helpful in locating this species. The only easily accessed location where it is possible to see the Hawai'i Creeper is at Kīpuka 21 along the Saddle Road. Government agencies have reintroduced the species into this forest oasis in recent years; nevertheless a sighting there remains unlikely. A more reliable option is a guided hike into the restricted Hakalau Forest National Wildlife Refuge.

Hawai'i Creeper, Oreomystis mana. (JACK JEFFREY)

Maui Creeper (Maui *'Alauahio*)

PAROREOMYZA MONTANA

Endemic
4.5 inches (11 cm)

Known also as the Maui Creeper, the Maui *'Alauahio* is found only on the high slopes of Haleakalā. When observing this small endemic honeycreeper, North American birders will be reminded of a warbler. Adult males are yellow green above and bright yellow below. Females are duller. The bill is short and straight. Unlike the bold black lores of the Maui *'Amakihi*, there is only a faint black smudge between the bill and the eye of the Maui *'Alauahio*. The bill and legs of both sexes are light pink, darker in immature birds. The Maui Creeper often travels in flocks that can be located by listening for its short, repeated "chip" call. The Maui *'Alauahio* rarely takes nectar. Birds are usually seen actively foraging among leaves and branches for insects. An easy place to see this species is in the *māmane* and *pūkiawe* scrub or in the tall introduced trees at Hosmer Grove.

Maui Creeper (Maui 'Alauahio*),* Paroreomyza montana. (JACK JEFFREY)

Maui Parrotbill

PSEUDONESTOR XANTHOPHRYS

ENDANGERED
Endemic
5.5 inches (14 cm)

This rare endemic honeycreeper inhabits the wet forest high on the windward slopes of Haleakalā on the island of Maui. The Maui Parrotbill forages in the subcanopy, where it uses its heavy hooked bill to crush dead and rotting branches in search of insect larvae. The sexes are fairly similar. They are yellow green above, lighter underneath. A broad yellow stripe above the eye extends from the base of the bill to the back of the head. The bill is pink overall, the top mandible darker. Legs and feet are black. The loud upslurred "chur-whee" call and the snap of branches are useful in locating the bird in the thick shrubs of the subcanopy. The only location where it is possible to see this specialized endemic forest bird is along the boardwalk in the Waikamoi Preserve, but access to this Nature Conservancy property is strictly prohibited without an authorized guide.

Maui Parrotbill, Pseudonestor xanthrophrys. (JACK JEFFREY)

'Apapane

HIMATIONE SANGUINEA SANGUINEA

Endemic
5 inches (13 cm)

'Apapane, Himatione sanguinea sanguinea *(juvenile)*. (JACK JEFFREY)

The *'Apapane* is Hawai'i's most common native forest bird. It is often seen flying high above the forest canopy in search of the brilliant red *lehua* flowers of the *'ōhi'a* tree. This Hawaiian honeycreeper is also a gifted singer. Biologists who study this beautiful endemic species are impressed with the almost endless variety of song it has to offer. The whirring noise the *'Apapane* makes with its wings also contributes to its repertoire. Although it prefers the nectar of the *lehua* blossom, it can occasionally be seen in the understory feeding on the nectar of other native and alien plants. The *'Apapane* is red with black wings and tail. The bill is black, short, and slightly decurved. The legs are also black. The belly and undertail coverts are white. Immature birds are brown above and pale white below. The *'Apapane* is a quick-moving bird and not easily approached. It is common at high elevations on all the main islands except Lāna'i, where it is rare.

'Apapane, Himatione sanguinea sanguinea. (JACK JEFFREY)

ʻIʻiwi

VESTIARIA COCCINEA

Endemic
6 inches (15 cm)

The *ʻIʻiwi,* extinct on Lānaʻi and nearly so on Molokaʻi and Oʻahu, is still relatively common in the high native forests of Kauaʻi, Maui, and the Big Island. The *ʻIʻiwi* feeds on a variety of nectar sources. The flowers of *māmane, ʻōhelo,* and *ʻākala* are a few, but the long salmon-colored bill is a perfect match for pollinating the curved blossoms of native lobelias. Immature birds are speckled with green, black, and yellow. Once valued for its bright red plumage, this handsome bird held an important role in Hawaiian culture. It is estimated that the feathers of 30,000 *ʻIʻiwi* were used to produce a single cape to adorn the *aliʻi* (chiefs). A collector on a visit to the Islands a century ago noted that native boys caught the *ʻIʻiwi* by hiding in the bushes with a curved flower held between thumb and forefinger. When the bird inserted its long curved bill, the boys pinched and held fast to their prize.

ʻIʻiwi, Vestiaria coccinea (*juvenile*). **(JACK JEFFREY)**

ʻIʻiwi, Vestiaria coccinea. **(JIM DENNY)**

Kīpuka 21, named for the adjacent mile marker on the Big Island's Saddle Road, is a wonderful roadside location for birds. Five of that island's endemic forest birds are common here. Even the endangered ʻAkiapōlāʻau *and Hawaiʻi Creeper have been reported in this oasis in the lava.* **(JIM DENNY)**

Crested Honeycreeper ('Ā'kohekohe)

PALMERIA DOLEI

ENDANGERED
Endemic
7 inches (18 cm)

The *'Ākohekohe,* or Crested Honeycreeper, is an endemic species restricted to the native forest high on the northeastern slope of Haleakalā on the island of Maui. In poor or backlit light, this odd honeycreeper looks very much like a large *'Apapane*. It is mostly black overall with silver feathers accenting the breast. The bill and legs are black. There is a large red patch behind the head, and a large orange ring encircles the eye. A prominent white crest adorns the head. The unusual song consists of croaks, buzzes, and whistles. It is fairly common in the Waikamoi Preserve, but access to this Nature Conservancy property is strictly prohibited without an authorized guide. There are rare reports of the *'Ākohekohe* feeding among the red *lehua* blossoms below the lookout railing at Hosmer Grove.

Crested Honeycreeper ('Ākohekohe), Palmeria dolei. (JACK JEFFREY)

Melodious Laughingthrush (Hwamei)

Alien
10 inches (25 cm)

GARRULAX CANORUS

The Hwamei is also called the Chinese Thrush or Melodious Laughingthrush. The Chinese immigrants who valued it as a cage bird probably brought it to Hawai'i at the turn of the twentieth century. This introduced species is well established in both lowland and mountain habitats on all islands with the exception of Lāna'i. Often heard but seldom seen, the species now thrives in areas of the native forest once occupied by native thrushes. The vocalizations of the Hwamei are loud, repeated phrases of couplets and triplets. The Hwamei is rusty brown with a yellow bill and a conspicuous white eye ring that extends into a white line behind the eye.

Melodious Laughingthrush (Hwamei), Garrulax canorus. (JIM DENNY)

Greater Necklaced Laughingthrush

GARRULAX PECTORALIS

Alien
13 inches (33 cm)

This large babbler, found only on the island of Kaua'i, was introduced from Southeast Asia perhaps as early as 1918. The Greater Necklaced Laughingthrush travels in small family groups that prefer thickly forested areas. Birds typically stay in a location for only a few days before moving elsewhere. This large, long-tailed bird is rusty brown above and white below. It has a white facial patch marked with a series of wavy black stripes extending behind the eye. It can easily be distinguished from the Hwamei by the dark lei or necklace that droops across the white breast. It is a difficult bird to find. Sightings have occurred in 'Ōma'o, Hulē'ia, Wailua Arboretum, and densely forested stream valleys in Kīlauea.

Greater Necklaced Laughingthrush, Garrulax pectoralis. (JIM DENNY)

White-rumped Shama

COPSYCHUS MALABARICUS

Alien
10 inches (25 cm)

This species was introduced to Hawai'i from Malaysia in 1931 and again in 1940. It is established on Kaua'i, O'ahu, Moloka'i, and possibly on Maui. The adult male is an attractive bird. It is glossy black except for a rust-colored belly and white rump. The black tail with white tips is longer than that of any other songbird in Hawai'i. Many consider the song of this species the most pleasing of all of Hawai'i's birdsongs. It is particularly vocal near sunset and at sunrise. Although shy in the forest, it can get quite habituated to people in backyard gardens. Raking leaves or mowing the yard will often entice birds to approach closer to pick up a disturbed insect or worm.

White-rumped Shama, Copsychus malabaricus. (JACK JEFFREY)

Japanese Bush-Warbler (Uguisu)

CETTIA DIPHONE

Alien
5.5 inches (14 cm)

This secretive bird was introduced to O'ahu from Japan in 1929. It is brownish above and beige below with a white superciliary stripe and long tail. The Japanese Bush-Warbler prefers to forage in dense underbrush and is often difficult to observe. The Japanese name for this bird is Uguisu, which is an imitation of its song. The song is powerful and makes one think that the bird is closer than it really is. It often begins with a long sustained note followed by one higher in pitch. It also utters a long, slow series of paired notes that can last for 20 seconds or more. The Japanese Bush-Warbler is now found throughout Hawai'i in both upland and lowland forests. On Moloka'i, it is the most common forest bird.

Japanese Bush-Warbler (Uguisu), Cettia diphone. (JIM DENNY)

Red-billed Leiothrix

LEIOTHRIX LUTEA

Alien
5.5 inches (14 cm)

Also known as the Peking Nightingale or Japanese Hill Robin, the multicolored Red-billed Leiothrix was introduced to Hawai'i from Asia early in the twentieth century. It now exists on all islands except Lāna'i and Kaua'i. The species was at one time present on Kaua'i, but for reasons unknown that population no longer exists. The plumage above gradually transitions from yellow on the head to black at the tip of the widely forked tail. It is yellow below, particularly bright at the throat. The red bill and red patch at the base of the primaries contribute to making this one of Hawai'i's most beautiful birds. The bird travels in small family groups that are more easily heard than seen.

Red-billed Leiothrix, Leiothrix lutea. (JIM DENNY)

Mariana Swiftlet

AERODRAMUS BARTSCHI

Alien
4 inches (10 cm)

The Mariana Swiftlet was introduced to O'ahu from Guam in 1962. It is also known as the Island, Gray, or Guam Swiftlet. This small bird is all dark with a slightly pale rump. The wings are short and pointed. The tail is slightly forked. In flight, it looks somewhat like a bat but with a more erratic flight. The species nests and roosts in an old irrigation tunnel, and possibly caves, high up in Hālawa Valley. It feeds mostly at dawn and dusk. Reports are rare, but birds have been seen flying high over the ridges at the head of the valley. The Mariana Swiftlet has been declared an endangered species on the island of Guam largely due to predation by the Brown Tree Snake.

Mariana Swiftlet, Aerodramus bartschi. (JACK JEFFREY)

Red Junglefowl *(Moa)*

GALLUS GALLUS

Alien
Male: 30 inches (76 cm)
Female: 17 inches (43 cm)

Brought by the Polynesians, the *Moa,* or Red Junglefowl, is Hawai‘i’s first introduced bird. The *Moa* is a bird mostly of the upland forest and is easily confused with the many chickens (*Gallus domesticus*) that roam the lowlands. As the name suggests, this species is red in contrast to the varied plumage of lowland chickens. Extensive interbreeding does occur, however. The birds are especially numerous on Kaua‘i. Look for them in the meadow fronting the Kōke‘e Museum. They are present by the dozens and are easily approached.

Red Junglefowl (Moa), Gallus gallus *(male).* (JIM DENNY)

Kalij Pheasant

LOPHURA LEUCOMELANOS

Alien
29 inches (73 cm)

Currently found only on the Big Island, this large chickenlike game bird was introduced to Hawai'i in 1962 from the southern foothills of the Himalayas. Adult males are an iridescent dark blue with a long gray breast and white barring on the back. Like the Ring-necked Pheasant, it has a large red facial patch surrounding the eye. The Kalij Pheasant is now common in many mesic to wet areas on the Big Island. It can often be seen early or late in the day as it forages in and near forested areas. Look for it at Kīpuka Puaulu, or Bird Park, located within Hawai'i Volcanoes National Park.

Kalij Pheasant, Lophura leucomelanos. (JACK JEFFREY)

Kalij Pheasant, Lophura leucomelanos *(female)*. (JACK JEFFREY)

The Kanahā Pond Wildlife Sanctuary is located alongside a busy highway less than a mile (0.7 km) from the bustling Kahului Airport on the island of Maui. Two species of endemic waterbirds are common here. Kanahā Pond and the Keālia National Wildlife Refuge, on Maui's south shore, are both important winter refuges for migrant ducks and shorebirds. (JIM DENNY)

WETLAND BIRDS ARE HAWAI'I'S LARGEST group of birds. Hawai'i's central location in the Pacific Ocean, nearly halfway between the continents of Asia and North America and midway between the Arctic breeding grounds and the islands of the South Pacific, has made these isolated islands a stopover point for migrant species of the Pacific rim. Eighty-one species of migrant shorebirds and waterfowl have been recorded in Hawai'i. Some, like the American Avocet and Stilt Sandpiper, are extremely rare, with only a single confirmed record for each species since the formal collection of data began. Others, like the Pacific Golden-Plover and Ruddy Turnstone, are frequent visitors on island lawns, parks, and meadows. In all, sixty-one migrants are illustrated here.

Also described in this group are six resident native wetland birds. Hawai'i's endemic wetland birds have fared better than native forest birds. In spite of predation by introduced mammals, every species of endemic wetland bird that was recorded here by the Europeans in 1778 can still be seen with relative ease in the taro patches, streams, and wildlife refuges around the state. This is especially true of Kaua'i, an island that lacks the predatory mongoose.

Pied-billed Grebe

PODILYMBUS PODICEPS

Indigenous
11 inches (28 cm)

Although widespread in North America, this small grebe is only rarely seen in Hawaiʻi. Breeding adult birds are brown with a black throat patch. The thick white bill has a black ring near the tip. Winter migrants seen in the Hawaiian Islands either have a faint ring or it may be lacking altogether. The Pied-billed Grebe often dives to escape danger and may surface far from the dive spot. There is a record of this species successfully breeding at ʻAimakapā Pond on the Big Island. The species bred at this site for about 20 years although numbers were small.

Pied-billed Grebe, Podilymbus podiceps. (JIM DENNY)

Great Blue Heron

ARDEA HERODIAS

Migrant
46 inches (117 cm)

This large heron is a rare straggler to the Hawaiian Islands. In Hawai'i it frequents the same habitat as the common Black-crowned Night-Heron but is nearly twice its size. This tall, lanky bird is mostly gray with whitish plumes over the back and breast. The face is white with a wide black stripe over the eye. The daggerlike bill is yellow. The long legs are gray. The Great Blue Heron usually feeds in shallow water, but like the ubiquitous Cattle Egret, it will also feed on land on insects and mice.

Great Blue Heron, Ardea herodias. (JACK JEFFREY)

Black-crowned Night-Heron (*'Auku'u*)

NYCTICORAX NYCTICORAX HOACTLI

Indigenous
25 inches (63 cm)

Named the *'Auku'u* by the Hawaiians, this night-heron is a solitary wading bird that often stands motionless on the banks of ponds and ditches while eyeing its prey in the water below. As the name Night-Heron implies, the *'Auku'u* is known to feed in both daylight and darkness. When flying overhead they sometimes utter a loud "kwok" noise. Immature birds are heavily streaked with brown and white. Breeding adults are adorned with a long, white head plume. The Black-crowned Night-Heron arrived in the Hawaiian Islands naturally, has bred here for centuries, and is thus considered a native bird; however, the Hawaiian form is as yet indistinguishable from the North American species.

Black-crowned Night-Heron ('Auku'u), Nycticorax nycticorax hoactli.
(JIM DENNY)

White-faced Ibis

PLEGADIS CHIHI

Migrant
23 inches (58 cm)

This tall, long-legged bird is a rare winter visitor to Hawai'i. The extremely long, curved bill and the iridescent purple and green plumage is impressive. All the birds that have been reported in the Islands have been juveniles or birds in nonbreeding plumage. This makes it very difficult to distinguish them from the very similar Glossy Ibis. The best key to separating the two is probably eye color. As the White-faced Ibis matures, the eye color changes from brown to red. The White-faced Ibis has been reported in the taro patches and wetlands of all the main islands with the exception of Lāna'i, which has very limited wetlands.

White-faced Ibis, Plegadis chihi. (JIM DENNY)

Fulvous Whistling-Duck

DENDROCYGNA BICOLOR

Indigenous
20 inches (51 cm)

The Fulvous Whistling-Duck, or Fulvous Tree-Duck, is yellowish brown on the face, belly, breast, and sides of the neck. It has a dark back and black wings. A broad black stripe extends from the crown down the back of the neck onto the upper back. The bill and legs are bluish gray. A white band can be seen at the base of the tail. The long neck, long legs, and upright posture easily set it apart from other ducks in Hawai'i. Like the Pied-billed Grebe, it dives for food. It is named for the whistlelike call that it gives in flight. The Fulvous Whistling-Duck is known to have successfully bred at Kahuku on the island of O'ahu, but the species has not been seen there in several years.

Fulvous Whistling-Duck, Dendrocygna bicolor. (JACK JEFFREY)

Greater White-fronted Goose

ANSER ALBIFRONS

Migrant
28 inches (71 cm)

This grayish brown barred goose is common in both North America and Eurasia. It is also a rare visitor to Hawai'i. The feet and legs are orange. The rump and the area under the tail are white. The Greater White-fronted Goose is named for the characteristic white band at the base of the pink bill. Although it can turn up in any suitable wetland in Hawai'i, the large ponds in Hilo town on the Big Island have been favored spots for this handsome goose.

Greater White-fronted Goose, Anser albifrons. (JIM DENNY)

Snow Goose

CHEN CAERULESCENS

Migrant
28 inches (71 cm)

It seems odd that any goose closely associated with the cold Arctic would choose to winter in balmy Hawai‘i. This mostly white bird is indeed a rare visitor to the Islands. Fewer than forty birds have ever been reported. The Kawai‘ele Sand Mine on the island of Kaua‘i was the winter home of this goose in 2006. The legs, bill, and feet are pink. The cutting edge of the bill has a dark lining that gives the bird its characteristic grin.

Snow Goose, Chen caerulescens. (JIM DENNY)

Canada Goose

BRANTA CANADENSIS

Migrant
45 inches (114 cm)

This large goose, so abundant and widespread on the North American continent, rarely visits the Hawaiian Islands. The Waiākea Pond in Hilo on the Big Island and the Hanalei National Wildlife Refuge on Kaua'i have been good places to look in recent years. It has a black head and a long black neck with a characteristic white band under the chin. The rump is also white. It feeds on both the water and the ground.

Canada Goose, Branta canadensis. (JIM DENNY)

Cackling Goose

BRANTA HUTCHINSII

Migrant
23 inches (64 cm)

Until recently, the Cackling Goose was considered a race of the Canada Goose. The two look very much alike but in fact are now regarded as separate species. Similar in size to the *Nēnē,* the Cackling Goose is about half the size of the Canada Goose. It has a shorter neck, shorter legs, a stubby bill, and a more stocky appearance. In Hawai'i, the Cackling Goose is a rare migrant. Most reports have come from Kaua'i and the Big Island.

Cackling Goose, Branta hutchinsii. (JIM DENNY)

Hawaiian Goose (*Nēnē*)

BRANTA SANDVICENSIS

ENDANGERED
Endemic
24 inches (61 cm)

The *Nēnē,* or Hawaiian Goose, is the state bird of Hawai'i. An endangered species, this endemic goose was hunted to the brink of extinction. Fossil records show that it was once abundant and widespread throughout Hawai'i, but by the early 1950s it was rarely seen in the wild. Today, through the efforts of successful captive propagation, the *Nēnē* once again nests in the wild on Kaua'i, Maui, Moloka'i, and the Big Island. The Kaua'i population is the most robust. The introduced Small Indian Mongoose is a serious nest predator of the *Nēnē.* Kaua'i is fortunate not to have this alien mammal. The *Nēnē* is smaller and has less webbing in the feet than most geese. Although it can occasionally be seen in the water, it prefers to feed on land.

Hawaiian Goose (Nēnē), Branta sandvicensis. (JACK JEFFREY)

Brant

BRANTA BERNICLA NIGRICANS

Migrant
28 inches (71 cm)

The Black Brant or Brent Goose, as it is sometimes called, is an Arctic sea goose that usually winters on the western seacoast of the North American continent. The Brant is also a rare visitor to Hawai'i. It feeds on marine grasses, so it is not surprising that most sightings occur on shallow reefs with an adequate supply of seaweed. This dark, stocky, short-necked bird has a diffuse white band that nearly encircles the neck.

Brant, Branta bernicla nigricans. (JIM DENNY)

Tundra Swan

CYGNUS COLUMBIANUS

Migrant
52 inches (132 cm)

This large, white bird is the most widespread swan in North America. It migrates tremendous distances in the winter months although rarely to the Hawaiian Islands. The Tundra Swan is all white with a long neck. The head and neck are often stained brown or red in the process of feeding. In most birds there is a yellow facial patch of various size between the black bill and eye.

Tundra Swan, Cygnus columbianus. (JIM DENNY)

Gadwall

ANAS STREPERA

Migrant
20 inches (51 cm)

At a distance, the adult male Gadwall is not a very impressive duck. Closer observation, however, reveals a delicately scalloped plumage that is quite stunning. Other features are a short black bill, orange legs, and a white speculum. A brown wash extends from the forehead over the crown and onto the back of the head. The Gadwall is a rare visitor to Hawai'i. Most reports come from Moloka'i.

Gadwall, Anas strepera. (JIM DENNY)

Eurasian Wigeon

ANAS PENELOPE

Migrant
20 inches (51 cm)

The Eurasian Wigeon is closely related to the American Wigeon. Although it is a less-common visitor than its American counterpart, it can occasionally be seen mixed in with flocks of American Wigeons in Hawai'i's wetlands. Males have a gray body and a chestnut head. A pale stripe accents the forehead. Aside from plumage, there are a couple of other small differences between the two species. The Eurasian Wigeon has longer wings and a proportionately smaller head. Hybrids of the two species are occasionally seen.

Eurasian Wigeon, Anas penelope. (ELAINE R. WILSON)

American Wigeon

ANAS AMERICANA

Migrant
19 inches (48 cm)

The American Wigeon is an abundant and widespread species in western North America. It is also one of the more common migrant ducks to winter in Hawai'i. In the Islands, Wigeons graze on the short grass that grows on the dikes around taro patches. They also feed on the surface of the water and often steal from diving ducks like Scaup or Ring-necked Ducks. The head of the male is a finely streaked gray. The crown is white. There is an attractive broad green stripe that extends from the eye to behind the head. In flight, a large white patch is visible on the forewing.

American Wigeon, Anas americana. (JACK JEFFREY)

Mallard

ANAS PLATYRHYNCHOS

Migrant
23 inches (58 cm)

The Mallard is abundant and widespread in North America but a rare migrant to Hawai'i. The male has an iridescent green head, chestnut breast, and a grayish body. A conspicuous white neck ring can be seen at the base of the neck. The female is plain by comparison and looks like the native *Koloa maoli,* or Hawaiian Duck. Both the male and female have a blue or purple speculum bordered in white. Legs of both sexes are orange. The Mallard is known as a puddle duck. It will feed in almost any small patch of water. A variety of domestic Mallard breeds have been imported to the Islands. These hybridized farm ducks are widespread and frequently hybridize with the endangered *Koloa maoli.*

Mallard, Anas platyrhynchos. (JIM DENNY)

Hawaiian Duck *(Koloa maoli)*

ANAS WYVILLIANA

ENDANGERED
Endemic
20 inches (51 cm)

The endemic Hawaiian Duck, or *Koloa maoli*, is closely related to the migratory Mallard and resembles the female of that species. In the Hawaiian species, there is very little difference in plumage between male and female; however, males are somewhat larger and have a darker head. In both sexes, the legs and feet are orange. Unfortunately, it is *Koloa*/Mallard hybrids that inhabit most islands. Kaua'i and the Kohala area of the Big Island are the only places where genetically pure Hawaiian ducks can still be seen. In the 1840s this duck was so plentiful that it was slaughtered by the hundreds to provision visiting whaling ships. Mammalian predators and habitat loss have reduced its numbers further. The Hawaiian Duck is listed as an Endangered Species by the U.S. Fish and Wildlife Service.

Hawaiian Duck (Koloa maoli), Anas wyvilliana. (JACK JEFFREY)

Blue-winged Teal

ANAS DISCORS

Migrant
15 inches (39 cm)

A resident of North America, this small duck is named for the conspicuous blue forewing that can be seen only in flight. The attractive adult male has a large, prominent, white crescent between the bill and the eye. The bill is a dark bluish black and the head dark charcoal black. The body is light brown and covered with black spots. Legs are yellow orange. Like all teal, it flies fast with a quick wing beat. The Blue-winged Teal is a rare winter migrant to Hawai'i.

Blue-winged Teal, Anas discors. (JIM DENNY)

Cinnamon Teal

ANAS CYANOPTERA

Migrant
16 inches (41 cm)

The Cinnamon Teal is similar in size and shape to the Blue-winged Teal. In breeding plumage, the adult male is a deep chestnut or cinnamon color. The speculum is green. The bill is dark and resembles that of the Northern Shoveler but is not as long or spatulate. The Cinnamon Teal is a common inhabitant of western North America but rarely migrates to the Hawaiian Islands. Most reports of this species are from Maui and Kaua'i.

Cinnamon Teal, Anas cyanoptera. (JIM DENNY)

Northern Shoveler (*Koloa mohā*)

ANAS CLYPEATA

Migrant
19 inches (48 cm)

Many species of ducks have been recorded in Hawai'i, but the Northern Shoveler winters in the Islands with such regularity that it too has been given a Hawaiian name, *Koloa mohā*. The word *mohā* (bright) describes the glossy green head of males in breeding plumage. The best way, however, to identify this species is not by color but by the shape and length of the bill. It is the only duck in which the bill, shaped like a wide spoon or a shovel, is longer than the head. The Northern Shoveler is a member of a group of ducks known as puddle or dabbling ducks. All the members of this group prefer shallow freshwater or brackish ponds, where they tip over, tail up, to grub the bottom for food.

Northern Shoveler (Koloa mohā), Anas clypeata. (JACK JEFFREY)

Northern Pintail (Koloa māpu)

ANAS ACUTA

Migrant
26 inches (66 cm)

Like the Northern Shoveler, this handsome waterfowl is one of the more common ducks to visit Hawai'i. Males are easily recognized by the prominent white stripe that extends up the neck and onto the head. Both sexes have pointed pinlike tails. The Northern Pintail, known to the Hawaiians as *Koloa māpu*, once migrated to Hawai'i by the thousands. Observers in the 1800s noted that there were large flocks of them on the lagoons and ponds of O'ahu and Kaua'i. One historian interprets the Hawaiian word *māpu* as "to rise or float off as a cloud," a depiction that must have adequately described hundreds of these ducks rising off the water to take flight.

Northern Pintail (Koloa māpu), *Anas acuta*. (JIM DENNY)

Green-winged Teal

ANAS CRECCA

Migrant
14 inches (37 cm)

At only 14 inches, the Green-winged Teal is the smallest of all the puddle or dabbling ducks. The adult male has a gray body with a red chestnut head. An attractive iridescent green band extends across the eye and down the side of the head. The black bill is narrow and short. A vertical white line can be seen in front of the wing. Green-winged Teals are very fast in the air. They fly in tight, fast groups that often turn quickly and erratically. The Eurasian Teal *(Anas crecca crecca)*, a separate race of the Green-winged Teal, also migrates to Hawai'i. The vertical white stripe on the body of the American species is absent in the Eurasian bird. Instead, a horizontal white stripe is present above the flanks. The Eurasian subspecies also has more white on the face. Both subspecies are rare visitors to Hawai'i. Both prefer to feed on mudflats but also frequent taro patches and irrigation ditches.

Green-winged Teal, Anas crecca. (ALAN D. WILSON)

Canvasback

AYTHYA VALISINERIA

Migrant
21 inches (53 cm)

At first glance this North American duck looks a lot like the Redhead but there are distinctive differences. The head slopes downward onto a long, black bill. The back and sides of the male are all white. Females are light brown with mottled gray sides. This diving duck is not a common visitor to Hawai'i; however, it has been reported from wetlands on the five larger islands. It prefers deeper water than other diving ducks. The sewage-treatment plants in Kona on the Big Island and at Kuilima on O'ahu have been favorite spots in recent years.

Canvasback, Aythya valisineria. (ALAN D. WILSON)

Redhead

AYTHYA AMERICANA

Migrant
19 inches (48 cm)

The adult male of this diving duck is aptly named. The rusty red head is round with a yellow eye. Like the Ring-necked Duck, the bill is tricolored. The bluish gray bill of both male and female is black at the tip. A narrow white ring separates the two colors. The breast is black, and the back and sides are a mottled gray. Although fairly common in mainland North America, the Redhead is a rare visitor to the Hawaiian Islands.

Redhead, Aythya americana. (JIM DENNY)

Ring-necked Duck

AYTHYA COLLARIS

Migrant
17 inches (43 cm)

From a distance this small black duck with white sides looks similar to a Scaup. Like the Lesser Scaup, the head of the adult male Ring-necked Duck is narrow and peaked, with a purplish gloss. Closely observed, however, separation of the two species is easy. Both sexes of the Ring-necked Duck have a white ring near the tip of the bill. The tip of the bill is the darkest portion of the bill. The male has an additional smaller white ring at the base of the bill. The Ring-necked Duck is an uncommon visitor to the Hawaiian Islands.

Ring-necked Duck, Aythya collaris. (JIM DENNY)

Tufted Duck

AYTHYA FULIGULA

Migrant
17 inches (43 cm)

This small diving duck is similar to the male Ring-necked Duck, but the back is darker and the sides whiter. A prominent black crest or tuft adorns the back of the head. The head has a purple gloss and a yellow eye. The blue gray bill has a black tip. Look for this uncommon visitor among flocks of Scaup and Ring-necked Ducks.

Tufted Duck, Aythya fuligula. (ALAN D. WILSON)

Greater Scaup

AYTHYA MARILA

Migrant
18 inches (46 cm)

Most of the Scaup that migrate to Hawai'i are Lesser Scaup, but a few Greater Scaup are recorded each season. When viewed at a distance the Lesser and Greater Scaup look nearly identical. Unfortunately, there is no one definitive feature that distinguishes the two. However, when several features are considered together, identification becomes easier. The Greater Scaup is bigger overall and, in relaxed birds, has a rounder, wider head. The bill is also slightly longer and heavier. The black button, or nail, at the tip of the bill is more pronounced. In flight, the white in the wing extends farther out toward the primaries than in the Lesser Scaup.

Greater Scaup, Aythya marila. (PETER LATOURETTE)

Lesser Scaup

AYTHYA AFFINIS

Migrant
16 inches (42 cm)

The Lesser Scaup is an abundant, widespread North American duck. It migrates in small numbers to the Hawaiian Islands and is consistently reported every year from wetlands around the state. The Lesser Scaup has a narrow black head with a high crown that appears angular or peaked. In good light the head of males often has a purple gloss. The small, thin bill has a small black button, or nail, at the tip. The sides are a mottled gray that appears all white or gray from a distance.

Lesser Scaup, Aythya affinis. (JIM DENNY)

Bufflehead

BUCEPHALA ALBEOLA

Migrant
14 inches (34 cm)

This small diving duck is fairly common and widespread on the North American continent but uncommon in Hawai'i. Males are a boldly colored white underneath, with a black back. There is a large white patch on the side of a greenish purple head. Most of the birds that show up in the Islands in the winter months are females. The female is dull black above and gray white below. The dark gray bill is short and stubby. The most distinguishing key is an elongated white patch on the side of a fuzzy black head. The Bufflehead is shy and prefers to stay far away from the shore when possible.

Bufflehead, Bucephala albeola *(female).* (ALAN D. WILSON)

Hooded Merganser

LOPHODYTES CUCULLATUS

Migrant
18 inches (46 cm)

All three North American species of mergansers have been seen in Hawai'i, but visits are rare. Nearly all have been females. Of the three species, the Hooded and Red-breasted Merganser *(Mergus serrator)* have been recorded the most. This small diving duck eats fish, crustaceans, and aquatic insects. The bill is long, thin, and serrated. The Hooded Merganser is the smallest of the three species and the only one with a yellow bill. Females are brownish above and white underneath. The head sports a prominent rusty crest. A diffuse white patch is located behind the eye.

Hooded Merganser, Lophodytes cucullatus *(female).* (ALAN D. WILSON)

Ruddy Duck

OXYURA JAMAICENSIS

Migrant
15 inches (36 cm)

This small, stocky duck is widespread in North America but is a rare migrant to the Hawaiian Islands. Ducks seen here in the winter months lack the characteristic bright blue bill seen in the breeding adult on the mainland. The Ruddy Duck is dark brown above and a streaked, light brown below. It has a large, prominent, white cheek patch. The flight pattern of this small duck is much like that of a coot. It flies low to the water, pattering its feet on the surface for a long distance before becoming airborne. It often dives to escape danger.

Ruddy Duck, Oxyura jamaicensis. (JIM DENNY)

Hawaiian Coot ('Alae ke'oke'o)

FULICA ALAI

ENDANGERED
Endemic
14 inches (36 cm)

The endemic Hawaiian Coot, or *'Alae ke'oke'o*, is descended from the American Coot (*Fulica americana*), a common, widespread, mainland species that, on extremely rare occasions, still migrates to the Hawaiian Islands. An endangered wetland inhabitant, the *'Alae ke'oke'o* is named for its white (*ke'oke'o*) frontal shield. Some local residents refer to it as the *'Alae* bird. This distinct Hawaiian native has a more extensive white shield than the American Coot. A few birds in Hawai'i retain the look of the ancestral species and demonstrate a white bill encircled near the tip with a narrow black band. A large red shield can also be seen in a small number of birds. Coots dive under water to feed and prefer shallow ponds and marshy areas. The Hawaiian Coot can be seen on all the Islands except Lāna'i, where it is rare.

Hawaiian Coot ('Alae ke'oke'o), Fulica alai. (JIM DENNY)

Hawaiian Moorhen ('Alae 'ula)

ENDANGERED
Endemic
13 inches (33 cm)

GALLINULA CHLOROPUS SANDVICENSIS

This endangered Hawaiian subspecies is descended from the Common Moorhen of North America. It is also known as the Mudhen, Gallinule, or, locally, as the *'Alae 'ula*. It gets its Hawaiian name from its red *('ula)* frontal shield. Birds are almost entirely black with extremely large yellow feet that have little or no webbing between the toes. The frontal shield and most of the bill are red. Only the tip of the bill is yellow. The Hawaiian Moorhen is a shy bird that prefers marshes and ponds with thick marginal vegetation. Introduced predatory mammals are a serious threat to this species. Although the Hawaiian Moorhen is common in suitable habitat on Kaua'i and to a lesser degree on O'ahu, it can no longer be found on the Big Island, Maui, Lāna'i, or Moloka'i, and attempts to reintroduce it to those islands have failed. In Hawaiian mythology, the *'Alae 'ula* was responsible for giving fire to the Hawaiian people.

Hawaiian Moorhen ('Alae 'ula), Gallinula chloropus sandvicensis. (JIM DENNY)

Hawaiian Stilt *(Ae'o)*

HIMANTOPUS MEXICANUS KNUDSENI

ENDANGERED
Endemic
16 inches (41 cm)

The *Ae'o* is a subspecies of the common Black-necked Stilt of North America. Like its mainland relative, this endemic stilt is black above and white below, with a long, needlelike black bill. The legs are bright pink. The Hawaiian race has more black on the neck and face. This species prefers to feed in shallow water or the muddy shore of ponds, although at times it will wade up to its belly to feed on aquatic arthropods and insects below or on the water surface. In winter months they can often be seen feeding singly or in small groups in fields flooded by heavy rains. The *Ae'o* noisily defends its nesting area and sometimes feigns injury to draw attention away from the nest. Mammalian predators and loss of habitat have reduced this endangered species to about 1,400 birds.

Hawaiian Stilt (Ae'o), Himantopus mexicanus knudseni. (JIM DENNY)

The winter rains that fill the Hanapēpē Salt Pond on the island of Kauaʻi create a good stopover point for migrant shorebirds. (JIM DENNY)

Black-bellied Plover

PLUVIALIS SQUATAROLA

Migrant
11.5 inches (29 cm)

This uncommon migratory species, also known as the Gray Plover, can be a difficult one to identify. In the winter months, a solitary Black-bellied Plover on a mudflat is so similar in appearance to the Pacific Golden-Plover *(Kōlea)* that it is hard to be sure. The discernment becomes easier if the two birds are standing next to each other. The Black-bellied Plover is slightly larger, stockier, with a thicker, more robust bill. The plumage is grayer overall than that of the *Kōlea*. In flight the Black-bellied Plover has black axillaries (armpits) and a white rump. The Black-bellied Plover, like the *Kōlea*, hunts by the run-and-peck method. Look for it in on mudflats or grassy fields near the coast throughout the Islands.

Black-bellied Plover, Pluvialis squatarola. (JIM DENNY)

Pacific Golden-Plover (*Kōlea*)

PLUVIALIS FULVA

Migrant
11 inches (28 cm)

The *Kōlea*, as local residents call it, is the most common winter migrant to Hawai'i. This species arrives in the Islands in August after spending May, June, and July nesting in the Arctic. It is a frequent sight on golf courses, parks, and other open spaces. Unlike many migratory shorebirds, the Pacific Golden-Plover is not limited to lowland coastal areas. It can even be seen along roads and other open areas in the high wet forest. The bird that arrives from northern latitudes in late summer looks very different from the one that leaves in the spring. By April, most have acquired a stunning black and white "tuxedo" look.

Pacific Golden-Plover, (Kōlea), Pluvialis fulva *(summer)*. (JACK JEFFREY)

Pacific Golden-Plover (Kōlea), Pluvialis fulva *(winter)*. (JACK JEFFREY)

Semipalmated Plover

CHARADRIUS SEMIPALMATUS

Migrant
7 inches (18 cm)

This small North American banded plover is an uncommon winter visitor to Hawai'i. It is named for the short webbing that connects the forward toes, a feature most observers are not likely to see. It is similar in appearance to another rare migrant, the Killdeer, but the Semipalmated Plover is smaller and has only a single dark breast band. This small plover has a short, black bill and yellow legs. Birds seen in Hawai'i in the winter months have brown facial markings and breast band. In nonbreeding plumage, the Semipalmated Plover is nearly identical to the Common Ringed Plover *(Charadrius hiaticula)*. There are only two confirmed records of the latter species in Hawai'i.

Semipalmated Plover, Charadrius semipalmatus. (JIM DENNY)

Killdeer

CHARADRIUS VOCIFERUS

Migrant
10.5 inches (27 cm)

This banded plover, although common and widespread on fields, parks, and lawns across North America, is a rare visitor to Hawai'i. Unlike the mainland, most Killdeer in Hawai'i are seen in wetland habitats. Adults are white underneath and gray brown above. The eye ring is a conspicuous bright orange. The most prominent feature, however, is the two bold black bands on the neck and breast. The Semipalmated Plover, another migrant shorebird to Hawai'i, is similar but smaller and with only one black band on the breast.

Killdeer, Charadrius vociferus. (JIM DENNY)

Greater Yellowlegs

TRINGA MELANOLEUCA

Migrant
14 inches (36 cm)

In plumage and leg color this winter visitor looks a lot like its smaller cousin, the Lesser Yellowlegs, but there are some notable differences. The Greater Yellowlegs is about 4 inches (10 cm) bigger and does not appear as delicate. Its bill is about 1.5 times the length of the head and is gray at the base. The bill is often slightly recurved. This bird is a rare migrant to Hawai'i. Most sightings have occurred on Maui, O'ahu, and Kaua'i.

Greater Yellowlegs, Tringa melanoleuca. (JIM DENNY)

Lesser Yellowlegs

TRINGA FLAVIPES

Migrant
10.5 inches (27 cm)

This yellow-legged shorebird is an uncommon visitor to the Islands. The Lesser Yellowlegs is about the same size as the more common Pacific Golden-Plover, but it has more of a delicate elongated shape with longer stiltlike legs. The bill is straight, thin, and black throughout its length. In comparison to the head, the bill is of an equal length. It is grayish brown above and white below. When disturbed it moves constantly in a nervous, agitated manner. A similar species, the Greater Yellowlegs, also migrates to the Islands but in far fewer numbers.

Lesser Yellowlegs, Tringa flavipes. (JIM DENNY)

Wandering Tattler (*'Ūlili*)

HETEROSCELUS INCANUS

Migrant
11 inches (28 cm)

The Wandering Tattler, as its name suggests, has an extensive migratory range that includes most of the tropical Pacific. It arrives in Hawai'i in late August and remains until April or May. The summer months are spent nesting in Alaska or Canada. Unlike the Pacific Golden-Plover, this species is rarely seen in grassy fields, preferring instead mudflats and exposed reefs, where it probes in the crevices of wave-washed rocks with its long, thin bill. The Hawaiian name for this species is *'Ūlili*. When disturbed it utters its name on takeoff—"uu-li-li-li." The popular Hawaiian song "'Ūlili ē" (author unknown) beautifully describes the voice and actions of this common visitor to Hawai'i.

Wandering Tattler ('Ūlili), Heteroscelus incanus. (JIM DENNY)

Spotted Sandpiper

ACTITIS MACULARIA

Migrant
7.5 inches (19 cm)

This small shorebird, widespread throughout the North American continent, is a rare migrant to the Hawaiian Islands. Most sightings have been on the island of O'ahu. The constant tail bobbing and distinctive flight pattern of the Spotted Sandpiper help to identify this bird. It typically flies close to the water with shallow wing beats interspersed with short glides. In breeding plumage, the Spotted Sandpiper is a beautiful bird, heavily spotted underneath with an orange bill. When seen in Hawai'i, this winter visitor is a dull gray brown above and all white below. The bill is dark, and a black stripe extends through the eye.

Spotted Sandpiper, Actitis macularia. (JIM DENNY)

Whimbrel

NUMENIUS PHAEOPUS

Migrant
17.5 inches (45 cm)

This curlew is a rare migrant to the Islands. At first glance, it is so similar in size and shape to the more common Bristle-thighed Curlew that some birds may go unreported. To distinguish the two, pay close attention to color and the call. The Whimbrel is darker overall, especially on the breast and neck. In flight, the rump lacks the cinnamon color seen in the Bristle-thighed Curlew. The lower mandible of the Whimbrel is lighter in color than the upper mandible for about half its length. On the Bristle-thighed Curlew, both mandibles are a similar color nearly to the tip of the bill. The most distinctive difference, however, is the call. The call of the Whimbrel is a loud "tee-tee-tee-tee-tee."

Whimbrel, Numenius phaeopus. (JIM DENNY)

Bristle-thighed Curlew (*Kioea*)

NUMENIUS TAHITIENSIS

Migrant
18 inches (46 cm)

The islands of the Pacific are the primary wintering ground for this majestic long-billed migrant. In the summer months it breeds on the remote tundra of northwestern Alaska, but as the scientific name suggests, it is capable of extremely long migratory flights even as far south as Tahiti. The Hawaiians named it the *Kioea*. The Bristle-thighed Curlew strongly resembles the Whimbrel, but the *Kioea* has a buffier rump and is more golden overall. The feathers are bristled at the thigh. The curve of the bill is slightly more pronounced. The call is a loud "chee uu whit." The Bristle-thighed Curlew is a shy bird that prefers secluded areas of short grass or undisturbed sand dunes. Moloka'i, South Point on the Big Island, and especially the Kahuku wetlands on O'ahu are good locations to look for this uncommon species.

The well-watered lawn of the Moloka'i Education Center is a frequent gathering place for the long-billed Bristle-thighed Curlew. (JIM DENNY)

Bristle-thighed Curlew (Kioea), Numenius tahitiensis. (JIM DENNY)

Marbled Godwit

LIMOSA FEDOA

Migrant
18 inches (46 cm)

Godwits are rare migrants to Hawai'i. Among the four species that have been recorded here, the Marbled Godwit and the Bar-tailed Godwit *(Limosa lapponica)* are the most frequent. All Godwits are large, long-legged shorebirds with bicolored bills. The bill is curved upward in most species. All Godwits feed by probing the bill into the mud or sand in a sewing-machine motion. Consult North American field guides for important keys used to distinguish these impressive long-distance migrants.

Marbled Godwit, Limosa fedoa (JIM DENNY)

Ruddy Turnstone (*'Akekeke*)

ARENARIA INTERPRES

Migrant
9.5 inches (24 cm)

Red legs and a black bib make it easy to distinguish this robust shorebird from Hawai'i's other winter migrants. In flight, the bold black-and-white pattern on the wings and back are eye-catching. After a summer of nesting in the Arctic tundra, the Ruddy Turnstone spends the months of August through April in Hawai'i and on coastal areas throughout the tropical Pacific. The Hawaiians named this bird the *'Akekeke*, a name that resembles its call. Unlike the solitary Pacific Golden-Plover, the Ruddy Turnstone prefers to feed in pairs or small groups. Ruddy Turnstones, as their name suggests, feed by flipping over small rocks with their bill to find the invertebrates that lie beneath. It is rarely seen on sandy beaches, preferring instead pebble-strewn shorelines, mudflats, or flooded pastures. It is a common winter visitor to Hawai'i.

Ruddy Turnstone ('Akekeke), Arenaria interpres. (JACK JEFFREY)

Red Knot

CALIDRIS CANUTUS

Migrant
10.5 inches (27 cm)

The Red Knot, or simply Knot as it is known in Europe, is a rare visitor to Hawai'i. The bird breeds in the tundra lands of the Northern Hemisphere. The North American race is in serious decline. Extensive hunting in the late 1800s and a reduction in a principal food source (Horseshoe Crab eggs) at important migratory stopover points on the east coast of the United States are blamed. When seen in nonbreeding plumage in Hawai'i the bird is a fairly drab gray above and white underneath. The black bill is medium in length, with a slight droop. It is similar in size to Dowitchers, with which it often associates. This species also feeds in the company of Sanderlings. Reports in Hawai'i have come from mudflats on the islands of Maui, Moloka'i, and O'ahu.

Red Knot, Calidris canutus. (JACK JEFFREY)

Sanderling *(Hunakai)*

CALIDRIS ALBA

Migrant
8 inches (20 cm)

The Sanderling is a joy to watch when seen on one of Hawai'i's many sandy beaches. It runs back and forth at the water's edge, actively probing with its bill for invertebrates hidden in the wet sand. The Hawaiian name for this common winter visitor, *Hunakai*, means "sea foam," an appropriate name for this perpetually moving shorebird. It is so adept at staying just ahead of the waves that it almost seems to be a part of the foam at the leading edge. Its little black legs move so fast that they are a blur. Some observers have described these movements as being like a windup toy. The *Hunakai* feeds both as an individual and in small flocks. After heavy rains, Sanderlings can be found even in flooded pastures. Birds seen in the Islands are pale gray above and white below. The shoulder has a dark patch. The legs and bill are black.

Sanderling (Hunakai), Calidris alba. (JACK JEFFREY)

Semipalmated Sandpiper

CALIDRIS PUSILLA

Migrant
6 inches (16 cm)

This species is one of a group of small, rare shorebirds similar in size and appearance that are commonly referred to as "peeps" or "stints." In winter plumage they can be difficult to separate. The Semipalmated Sandpiper resembles the Sanderling but is smaller. The feet are partially webbed, but this feature is hard to discern when the bird is wading in the water or standing in the mud. This rare visitor has dark legs and a short, straight, tubular-shaped bill that is somewhat blunt at the tip.

Semipalmated Sandpiper, Calidris pusilla. (JIM DENNY)

Western Sandpiper

CALIDRIS MAURI

Migrant
6.5 inches (17 cm)

Once considered a regular migrant, this small shorebird is an increasingly rare visitor to the Islands. It is a half-inch (1.3 cm) larger than other "peeps." The legs and bill are black. The bill is longer than that of the Semipalmated Sandpiper and slightly drooped. Most adult migrants are uniformly gray. Another member of the group that has been recorded in Hawai'i is an Asian species, the Red-necked Stint *(Calidris ruficollis)*. Refer to North American field guides for clues to identifying this and other species in this group.

Western Sandpiper, Calidris mauri. (JACK JEFFREY)

Least Sandpiper

CALIDRIS MINUTILLA

Migrant
6 inches (15 cm)

The Least Sandpiper is an uncommon visitor to the Islands. Every year a half-dozen or so are reported from various wetlands around the state. Of all the "peeps" it is the one most often reported in Hawai'i and the only one of the group with yellow legs. This is a characteristic that might be missed if the legs are coated with mud. The upper breast is streaked with brown. The bill is thin and black with a slight downward droop at the tip.

Least Sandpiper, Calidris minutilla. (JIM DENNY)

Pectoral Sandpiper

CALIDRIS MELANOTOS

Migrant
9 inches (22 cm)

Pectoral Sandpipers usually spend the winter months in South America. A few migrate to Hawai'i. Look for this uncommon bird in the months of August, September, and October on mudflats and shallow ponds. The Pectoral Sandpiper and the Sharp-tailed Sandpiper are very similar in size and appearance. To distinguish the two, pay attention to both the breast and the chestnut cap. The rusty crown of the Pectoral Sandpiper does not contrast markedly with the rest of the head plumage. The conspicuous streaks on the breast do, however, contrast strongly with the white, non-streaked underbelly. There is a definite demarcation between the two areas on the underside of this species.

Pectoral Sandpiper, Calidris melanotos. (JIM DENNY)

Sharp-tailed Sandpiper

CALIDRIS ACUMINATA

Migrant
8.5 inches (22 cm)

Like the Pectoral Sandpiper, this uncommon species uses the Hawaiian Islands as a fall stopover on its way south. The Sharp-tailed Sandpiper breeds in Siberia in the summer months and winters in Australia. It usually arrives a little later than the Pectoral Sandpiper. Look for it in September and October on mudflats, shallow ponds, and other freshwater wetlands. The Sharp-tailed Sandpiper and the Pectoral Sandpiper are very similar in appearance. To distinguish the two, pay attention to both the breast and the chestnut cap. The rusty crown of the Sharp-tailed Sandpiper is brighter and contrasts markedly with the rest of the head plumage. There is a prominent white superciliary stripe that contributes to this appearance. The streaks on the breast do not, however, contrast strongly with the white, nonstreaked underbelly. The two areas diffuse into one another.

Sharp-tailed Sandpiper, Calidris acuminata. (JIM DENNY)

Dunlin

CALIDRIS ALPINA

Migrant
8.5 inches (22 cm)

Although the Dunlin is one of the most common shorebirds on the coasts of North America, only a few make the long journey across the Pacific to Hawai‘i. Birds seen here in the fall are a plain grayish brown above and white below. The key to identifying the Dunlin is the narrow black bill that droops at the tip. The Dunlin is an active bird that feeds in shallow water as well as on mudflats and marshy shorelines.

Dunlin, Calidris alpina. (PETER LATOURETTE)

Curlew Sandpiper

CALIDRIS FERRUGINEA

Migrant
8.5 inches (22 cm)

This Eurasian species is a rare visitor to the Hawaiian Islands but is so similar in size and appearance to the more common Dunlin that the species may be underreported. In winter plumage both birds are gray above and white below. The Curlew Sandpiper has a slightly longer neck and legs, but the most distinguishing features are the white rump and bill shape. The bill of the Curlew Sandpiper has a downward curve that extends the entire length of the bill. The bill narrows to a fine tip. The bill of the Dunlin also becomes slimmer toward the end, but it is thicker throughout and is drooped mostly at the tip.

Curlew Sandpiper, Calidris ferruginea. (MICHAEL WALTHER/O'AHU NATURE TOURS)

Ruff

PHILOMACHUS PUGNAX

Migrant
10–12 inches
(25–31 cm)

When seen foraging on grassy lawns in Hawai'i, the Ruff looks like a long-legged, long-necked *Kōlea* or Pacific Golden-Plover. A closer look at this uncommon winter migrant reveals a longer, thinner bill and a head that appears disproportionately small compared with the stocky body. The Ruff is a Eurasian species. Like the *Kōlea*, it feeds on mudflats and shallow ponds as well as on schoolyards, soccer fields, and pastures. The female of the species is called a Reeve. It is similar to the male but smaller.

Ruff, Philomachus pugnax. (JIM DENNY)

Short-billed Dowitcher

LIMNODROMUS GRISEUS

Migrant
11 inches (28 cm)

The Short-billed Dowitcher is a rare visitor to Hawai'i. Some may go unreported because in the winter months, when migrant shorebirds are in nonbreeding plumage, it is extremely difficult to tell the difference between it and the more common Long-billed Dowitcher. As the name implies, some birds do have shorter bills, but there is often enough overlap in bill length to make them two of the hardest birds to separate in the field. The Short-billed Dowitcher is lighter gray with fine streaks on the face. The breast is finely speckled. Even the most competent observers, however, are sometimes confused. Seasoned birders agree that the best way to distinguish the two is the call. Long-billed Dowitchers are talkative when feeding in flocks and utter a "keek-keek-keek" call when taking flight. Short-billed Dowitchers are usually silent when feeding and utter a mellow "tuu-tuu-tuu" call when taking flight.

Short-billed Dowitcher, Limnodromus griseus. (JACK JEFFREY)

Long-billed Dowitcher

LIMNODROMUS SCOLOPACEUS

Migrant
11.5 inches (29 cm)

The impressive long bill of this uncommon winter migrant sets it apart from other more common visitors like the Pacific Golden-Plover, Ruddy Turnstone, and Sanderling. It may be confused with the Wandering Tattler *('Ūlili)* that also has a long bill and yellow legs, but the feeding habits of the two species are very different. The Long-billed Dowitcher feeds by probing with its bill in a manner that has been described as like the rapid up-and-down movement of a sewing machine needle. Look for it on mudflats and along the shorelines of freshwater ponds and taro patches.

Long-billed Dowitcher, Limnodromus scolopaceus. (JIM DENNY)

Wilson's Snipe

GALLINAGO DELICATA

Migrant
10 inches (26 cm)

The Wilson's Snipe is a rare winter visitor to Hawai'i. The bird prefers boggy habitats or flooded pastures, where it probes the mud for food. It is a master at concealment. Unless the bird moves, it is quite difficult to locate. The size and shape of the bill may lead to confusion with the Long-billed Dowitcher, another winter migrant, but the Wilson's Snipe is stockier with bold white stripes on the head. The feeding habits are also different. The Wilson's Snipe is nearly always seen near or in vegetation, whereas the Long-billed Dowitcher feeds in more exposed areas.

Wilson's Snipe, Gallinago delicata. (JIM DENNY)

Wilson's Phalarope

PHALAROPUS TRICOLOR

Migrant
9 inches (24 cm)

Phalaropes are rarely seen in the Hawaiian Islands. Among the three species that have been recorded here, the Wilson's Phalarope and the Red Phalarope *(Phalaropus fulicarius)* occur with the greatest frequency. A North American genus, this rare winter visitor is unusual among shorebirds because of the manner in which it feeds. Although it does wade along the shore and in shallow water like other shorebirds, the feet of the phalarope are partially lobed, a characteristic that allows it to swim. When feeding on the water, they spin in tight circles, creating a vortex that stirs up food from the mud beneath them. The Wilson's Phalarope has a long, thin, needle-like bill. Most birds seen in the winter months in Hawai'i are gray above and white below. The face is white with a gray eye stripe.

Wilson's Phalarope, Phalaropus tricolor. (JIM DENNY)

Pomarine Jaeger

STERCORARIUS POMARINUS

Migrant
21 inches (53 cm)

The Pomarine Jaeger, or Pomarine Skua, breeds in the Arctic. In the winter months it can occasionally be seen at sea around the main Hawaiian Islands. This large bird is dark brown above with broad wings. The Pomarine Jaeger is polymorphic. Some are dark below, some white. It has a thick neck that gives it a pot-bellied appearance. Like the *'Iwa,* Jaegers steal food from other birds at sea. They are highly maneuverable and relentless in pursuit of their victim. The species has been reported regularly in Māmala Bay on O'ahu. Other locations are Kīlauea Point on Kaua'i and from South Point on the Big Island. Chances for a sighting improve farther from shore. A pelagic trip will often turn up this bird.

Pomarine Jaeger, Stercorarius pomarinus. (MICHAEL WALTHER/O'AHU NATURE TOURS)

Laughing Gull

LEUCOPHAEUS ATRICILLA

Migrant
16 inches (38 cm)

Gulls are not common in Hawai'i. Most sightings occur in the winter months, a time when these stragglers lack the characteristic breeding plumage. In the summer months, on the eastern coast of North and Central America, the adult Laughing Gull has a distinctive black hood and a dark red bill. Birds seen in the winter months in Hawai'i are drab by comparison. In addition most gulls that visit the Islands are immature, a fact that further complicates identification. To accurately identify any gull, pay close attention to details of bill color and markings, leg color, and plumage in flight. The Laughing Gull has been reported on all the Islands at harbors and at wetlands near the seashore.

Laughing Gull, Leucophaeus atricilla. (JIM DENNY)

Bonaparte's Gull

CHROICOCEPHALUS PHILADELPHIA

Migrant
13 inches (33 cm)

In summer months, this gull is a common sight on both east and west coasts of the United States and Mexico. It can also be seen around the Great Lakes. It rarely wanders to Hawai'i. Most stragglers to the Islands are reported from mid- to late winter. Birds seen here lack their characteristic breeding plumage. The Bonaparte's Gull is small with graceful ternlike flight. The legs are pink and the bill black. It is nearly all white except for black trailing edges of the wings and tail. It also has short, narrow, black wing bars and a black patch behind the eye. Any small gull sighted in the Islands is likely to be this species. Other gulls that have been reported in Hawai'i are the California Gull *(Larus californicus)*, Franklin Gull *(Larus pipixcan)*, Herring Gull *(Larus argentatus)*, and Western Gull *(Larus occidentalis)*.

Bonaparte's Gull, Chroicocephalus philadelphia. (PETER LATOURETTE)

Ring-billed Gull

LARUS DELAWARENSIS

Migrant
17 inches (43 cm)

Gulls prefer broad, shallow tidal zones, conditions not found on tropical islands. This fact helps to explain why few gulls occur in the Hawaiian Islands. One of the more regular of these uncommon stragglers to Hawai'i is the Ring-billed Gull. This species is abundant along North American coastlines. It progresses through three years of different plumage before attaining its characteristic adult plumage, so when seen in winter months in the Islands it can be difficult to identify. Most Ring-billed Gulls that arrive in Hawai'i are first- or second-winter birds. The diagnostic feature for adults of this species is a black ring near the tip of the bill.

Ring-billed Gull, Larus delawarensis. (JIM DENNY)

Glaucous-winged Gull

LARUS GLAUCESCENS

Migrant
26 inches (66 cm)

An inhabitant of the western coast of North America, this large stocky gull is another uncommon wanderer to the Hawaiian Islands. The mantle is uniformly pale gray. It is the only large gull in which the color of the primaries matches that of the mantle. The tips of the primaries are white. The yellow bill is large and stout with a red spot near the tip of the lower mandible. The legs are pink and the eyes black. Identifying large gulls can be difficult. It may be helpful to remember that the Glaucous Gull, a less-frequent wanderer to Hawai'i, is all white (no gray mantle). All other mature large gulls have some degree of black in the primary wing tips. Most Glaucous-winged Gulls that straggle to the Islands do so in the winter and spring, and most are juveniles. To accurately identify this gull in all of its plumages, the reader should consult North American field guides.

Glaucous-winged Gull, Larus glaucescens. (JIM DENNY)

Glaucous Gull

LARUS HYPERBOREUS

Migrant
22–29 inches
(56–74 cm)

This large white gull breeds on the extreme northern coast of Alaska and Canada and migrates down both the west and east coasts of North America in winter months. It is considered scarce in the United States and is a rare visitor to the Hawaiian Islands. It is similar in size and appearance to the more common Glaucous-winged Gull, and the two are occasionally confused. The Glaucous Gull is pale gray above and all white below. There are no black or dark gray marks on the wing tips. When viewed from underneath, the primaries of the Glaucous Gull appear translucent. Most gulls seen in Hawai'i in late winter and early spring are immature. Juvenile birds have a pink bill with black tip. The photo of this species was taken at the Kawai'ele Sand Mine on leeward Kaua'i, a popular location in recent years for visiting gulls.

Glaucous Gull, Larus hyperboreus. (JIM DENNY)

The Kīlauea Point National Wildlife Refuge on the island of Kaua'i is an outstanding location for seabirds. It is not uncommon for graceful seabirds to soar within mere feet of visitors to this historic lighthouse. (JIM DENNY)

SEABIRDS ARE NOT ONLY INSPIRING to watch as they gracefully maneuver on the air currents above the ocean surface, but many species are valuable assets to local fishermen, who depend on these birds to locate the large *'Ahi*, or Yellow-fin Tuna, that frequent Hawaiian waters. Large, swirling gatherings of mixed-species seabird flocks congregate over schools of fish near the surface. The *'Ahi* pursue the same fish from below.

Short of being in the middle of one of these "bird piles," as local residents call them, there is no better place in the Hawaiian Islands to view them up close than the Kīlauea Point National Wildlife Refuge. Of all the species of seabirds that are known to frequent the Hawaiian Islands, eight can regularly be seen at this scenic peninsula. In all, twenty-two species of seabirds are illustrated in this section.

The population of seabirds on and around the main Hawaiian Islands is greatly reduced from numbers before human settlement. Introduced feral cats, dogs, rats, and mongooses are serious predators on nesting birds. Most breeding colonies are now restricted to predator-free offshore islets. Ka'ula and Lehua islets, both west of Kaua'i, and Mokumanu and Mānana islets, off the windward coast of O'ahu, are four of the most important breeding sites for Hawaiian seabirds.

Laysan Albatross (*Mōlī*)

PHOEBASTRIA IMMUTABILIS

Indigenous
32 inches (81 cm)

With a wingspan of nearly 6 feet (2 m), the *Mōlī* is unmistakable in the air as it effortlessly glides on the winds near the surface of the ocean. Its clumsiness on land, however, has led some to call it the gooney bird. On Kaua'i, from November to July, birds nest at the Kīlauea Point National Wildlife Refuge, in the surrounding Princeville area, and at the Pacific Missile Range Facility on the leeward coast. They also nest at Ka'ena Point on O'ahu. Reports of the Laysan Albatross are increasing on the islands of Moloka'i, Lāna'i, and Hawai'i. It is a mystery how this magnificent indigenous seabird can return to nest, not only on the same island, but also to the exact spot from which it hatched. The *Mōlī* wanders widely across the North Pacific in search of food. Two thousand miles (3,200 km) is not uncommon.

Laysan Albatross (Mōlī), Phoebastria immutabilis. (JACK JEFFREY)

Black-footed Albatross

PHOEBASTRIA NIGRIPES

Indigenous
32 inches (81 cm)

This large seabird is dark brown throughout with a small amount of white at the base of the bill and behind the eye. The legs and bill are dark. At sea, it glides very close to the surface of the water with stiff outstretched wings. Like the Laysan Albatross, the Black-footed Albatross also visits the main islands but in fewer numbers. It has been reported from the same locations as the Laysan Albatross. Up to twenty pairs nest on Lehua Islet. The Black-footed Albatross occasionally follows boats at sea. A pelagic trip off of Kaua'i is the most likely way to find this species. Birds are usually seen between the months of November and March.

Black-footed Albatross, Phoebastria nigripes. (JACK JEFFREY)

Hawaiian Petrel (*'Ua'u*)

PTERODROMA SANDWICHENSIS

ENDANGERED
Endemic
17 inches (43 cm)

This uncommon seabird was named *'Ua'u* by the Hawaiians for the moaning "uu-aa-uu" sound emanating from its burrow at night during nesting season. This call is also given in flight. An endemic species, it is listed as endangered by the U.S. Fish and Wildlife Service. Few birds remain because feral cats and the alien mongoose prey upon the bird at the nest. Until recently, the most well-known colony was at the summit of Haleakalā, but a new, rather large colony has been discovered on the high slopes of Lāna'i. The bird is very dark above and white below. The white extends up onto the forehead. The wings are narrow and long. When seen flying overhead at dusk on its way inland, it looks very much like a very large bat. This large seabird is strictly pelagic, and when not nesting, it is rarely seen from shore.

Hawaiian Petrel ('Ua'u), Pterodroma sandwichensis. (JIM DENNY)

Bulwer's Petrel (*'Ou*)

BULWERIA BULWERII

Indigenous
11 inches (27 cm)

The Bulwer's Petrel, or *'Ou,* is a small, dark brown petrel with long, pointed wings and a long, tapered tail. When seen in flight, the wings have a narrow light brown diagonal bar on the upper surface. At sea, it often keeps company with Wedge-tailed Shearwaters. Its erratic flight is close to the water and consists of short wing beats interspersed with glides. The Bulwer's Petrel is a regular sight in the waters around Hawai'i but difficult to see from shore. Although it nests on a few of the offshore islets visible from the coast, it generally only comes to these nest locations after dark. The best time of year to try for a glimpse of this small seabird is between April and October. There are rare reports of the *'Ou* at dusk off Kīlauea Point on Kaua'i and near Mānana Island on O'ahu.

Bulwer's Petrel ('Ou), Bulweria bulwerii. (JIM DENNY)

Wedge-tailed Shearwater ('Ua'u kani)

Indigenous
17 inches (43 cm)

PUFFINUS PACIFICUS CHLORORHYNCHUS

Named *'Ua'u kani* by the Hawaiians for its ghostly burrow call, this seabird is abundant and widespread in the tropical Pacific. Local fishermen know it as the *Aku* bird. In flight, it wanders from side to side over the water and makes high banking turns in the wind. The name, "shearwater," was coined by English sailors. To them it seemed that the bird flew so close to the water that the surface was cut or sheared. This indigenous Hawaiian species can occasionally be seen in flight from coastal headlands throughout Hawai'i, but the easiest place to see them at their burrows is at the Kīlauea Point National Wildlife Refuge on Kaua'i. From March through November, nesting birds can be seen by the dozens in their burrows alongside the pathway to the lighthouse. The Wedge-tailed Shearwater is brown above and gray white below. The colors melt into one another and are not contrasting as in the Newell's Shearwater.

Wedge-tailed Shearwater ('Ua'u kani), Puffinus pacificus chlororhynchus. (JIM DENNY)

Christmas Shearwater

PUFFINUS NATIVITATIS

Indigenous
14 inches (35 cm)

This shearwater is named for Christmas Island, an atoll near the equator discovered on Christmas Eve, 1777, by Captain James Cook. Thousands of these sooty black seabirds nest there. A few also nest on offshore islets around the main Hawaiian Islands. It is similar to the more common Wedge-tailed Shearwater but smaller with a shorter, rounded tail. In flight, the wing beats are stiff, rapid, and interspersed with long glides. On Kaua'i, birds have been seen at dusk flying near Kīlauea Point National Wildlife Refuge or flying inland just east of 'Ele'ele. On O'ahu, reports have come from ocean viewpoints east of Hanauma Bay. A pelagic trip greatly increases the chance of seeing this small all-dark shearwater.

Christmas Shearwater, Puffinus nativitatis. (JACK JEFFREY)

Newell's Shearwater (*'A'o*)

PUFFINUS AURICULARIS NEWELLI

THREATENED
Endemic
13 inches (33 cm)

The name, *'A'o*, given to this seabird by the Hawaiians, is an accurate description of its flight call. At night when traveling inland to its mountain burrows, it rapidly and repeatedly calls out, "Ah-oh, ah-oh, ah-oh, ah-oh," a sound reminiscent of a braying jackass. This endemic subspecies of the Townsend's Shearwater is not common in Hawaiian waters and is considered "threatened" by the U.S. Fish and Wildlife Service. It is smaller than the more common Wedge-tailed Shearwater, whiter underneath, and darker above with colors that are more contrasting than those seen in the Wedge-tailed Shearwater. It flies very close to the water in what has been described as a "flutter, flutter, glide–flutter, flutter, glide" pattern. The Newell's Shearwater nests on Kaua'i and the Big Island and possibly on the other large islands as well. On Kaua'i, a Newell's Shearwater fallout is an annual event. Each fall hundreds of birds become disoriented by the glare of street and park lights and fly around them until exhausted. These birds eventually fall to the ground, where they are often killed by cats and dogs or run over by automobiles.

Newell's Shearwater ('A'o), Puffinus auricularis newelli. (JIM DENNY)

Band-rumped Storm-Petrel ('Akē'akē)

OCEANODROMA CASTRO

Indigenous
9 inches (23 cm)

The Hawaiians named this small seabird the *'Akē'akē* for its call. In the evening hours as it makes its way inland to its nest burrow high in the mountains it utters a call that sounds a lot like someone rapidly rubbing a finger on wet glass. It is thought to nest on Kaua'i, the Big Island, and possibly Maui. The *'Akē'akē* has a broad, pure white band across the rump, and faint dark brown dorsal wing bars. The tail is squarish or slightly forked. When seen at sea, the flight is more like that of a shearwater than a storm-petrel. Like the Hawaiian Petrel, the species is rarely seen from shore. Pelagic trips on island waters occasionally report this small bird. The state of Hawai'i has listed it as an endangered species.

Band-rumped Storm-Petrel ('Akē'akē), Oceanodroma castro. (JIM DENNY)

White-tailed Tropicbird (Koaʻe kea)

PHAETHON LEPTURUS DOROTHEAE

Indigenous
27 inches (69 cm)

The indigenous *Koaʻe kea* can often be seen soaring in canyons, dormant calderas, or in the vicinity of rocky cliffs throughout the Hawaiian Islands. Between March and October the bird nests from coastal areas to inland valleys, usually in rocky crevices. This choice of nest site affords some protection from alien rats, cats, and mongoose that have decimated many other species of Hawaiian seabirds. Adult birds are white with a pair of long central tail feathers that often appear as one. Black wing bars are visible when viewed from above. When seen at sea the White-tailed Tropicbird has a direct, graceful, constantly flapping flight with very little gliding.

White-tailed Tropicbird (Koaʻe kea), Phaethon lepturus dorotheae.
(JIM DENNY)

Red-billed Tropicbird

PHAETHON AETHEREUS

Indigenous
18 inches (45 cm)

This beautiful seabird is the least numerous of the world's three species of tropicbirds. It is a rare visitor to the Hawaiian Islands. At a glance, it looks much like a White-tailed Tropicbird or Red-tailed Tropicbird and may go unnoticed. The tail is long and white and the upper parts darkly barred. The bill is a bright red. When searching among the more common tropicbirds seen in Hawai'i, look for the distinctive combination of red bill and long white tail streamers. In recent years, the Red-billed Tropicbird has been making near-annual appearances at Hālona Blowhole on O'ahu and at the Kīlauea Point National Wildlife Refuge on Kaua'i.

Red-billed Tropicbird, Phaethon aethereus. (MICHAEL WALTHER/O'AHU NATURE TOURS)

Red-tailed Tropicbird (Koa'e 'ula)

PHAETHON RUBRICAUDA

Indigenous
37 inches (93 cm)

Named for its long, red tail feathers, the *Koa'e 'ula* is delightful to watch during the mating season of March to October. In an elaborate courtship display, birds fly backward, as if rowing a boat, and at times the bird is nearly stationary on the wind. From a distance the beautiful red streamers are hard to see, giving it a tailless look. The Red-tailed Tropicbird nests under vegetation or on cliffs near the coast, and although graceful in the air it is so awkward on land that it cannot stand up or walk forward without falling. Like boobies, tropicbirds dive headlong into the water from great heights and are known to go as deep as 10 feet (3 m) to catch fish. The *Koa'e 'ula* is not common on all the islands. It can be seen with ease only from March to September at the Kīlauea Point National Wildlife Refuge on Kaua'i and near the Hālona Blowhole on O'ahu.

Red-tailed Tropicbird (Koa'e 'ula), Phaethon rubricauda. (JACK JEFFREY)

Masked Booby (ʻĀ)

SULA DACTYLATRA PERSONATA

Indigenous
32 inches (81 cm)

The Masked Booby is the largest of the three species of boobies present in the main islands, but it exists here in far fewer numbers than the other two species. On rare occasions, this impressive seabird can be seen cruising by the lighthouse at the Kīlauea Point National Wildlife Refuge on Kauaʻi or near Mānana Island, an islet off the northeastern coast of Oʻahu. Pelagic trips often report them near Mokumanu Islet. A few are known to nest on this small islet off the windward side of Oʻahu. When seen in flight, it resembles a Red-footed Booby, but the Masked Booby is larger with a black tail, and it flies higher above the water. The Masked Booby has an extensive black mask surrounding the eye and on the face at the base of the bill.

Masked Booby (ʻĀ), Sula dactylatra personata. (JACK JEFFREY)

Brown Booby (ʻĀ)

SULA LEUCOGASTER PLOTUS

Indigenous
30 inches (76 cm)

The Hawaiians apparently did not distinguish the three species of boobies that frequent the coastal waters of Hawaiʻi because they gave them all the same name, *ʻĀ*. The unflattering name, Booby, is from English sailors who considered these birds stupid because they were so easily grabbed when landing on the ship's rigging. The Brown Booby has a strong steady flight interspersed with frequent glides. At a distance, it appears all dark until it banks on the wind to reveal its all-white underside. Like the Red-footed Booby, it will dive from great heights to catch a fish. Typically, a single bird is seen cruising along the nearshore waters in search of prey. When a large school of fish is present, however, dozens of Brown Boobies gather to feed as a group.

Brown Booby (ʻĀ), Sula leucogaster plotus. (JACK JEFFREY)

Red-footed Booby ('Ā)

SULA SULA RUBRIPES

Indigenous
28 inches (71 cm)

The Red-footed Booby, or *'Ā*, is the most common of the three species of *Sula* in Hawai'i. At sea it often flies close to the water in a **V**-like formation with others of its species. It is a very capable fisher, sometimes plunging into the ocean from heights of up to 100 feet (30 m) to pursue its prey. The best place to see them up close in Hawai'i is at the Kīlauea Point National Wildlife Refuge on Kaua'i. They nest by the hundreds on the hillside adjacent to the lighthouse. It builds a nest atop short bushes, a behavior that helps protect it from mammalian predators. When the Red-footed Booby stalls on the wind to land, the sun shining through the bright red feet is a beautiful sight.

Red-footed Booby ('Ā), Sula sula rubripes. (JACK JEFFREY)

Great Frigatebird (*'Iwa*)

FREGATA MINOR PALMERSTONI

Indigenous
43 inches (109 cm)

The Great Frigatebird is an unparalleled flying machine because of its ability to outmaneuver other seabirds and for its capacity to glide for hours on warm updrafts. With a 7-foot (2.1 m) wingspan, this forked-tailed giant can easily soar as high as 500 feet (150 m)—higher than any other seabird. *'Iwa*, the Hawaiian name for this species, means "thief." The Great Frigatebird is adept at stealing food from boobies, shearwaters, and tropicbirds as they return with food to the nest. From great heights, they swoop down to harass their victims until they disgorge their meal. Sailors of old referred to them as "man-o'-war birds" because they reminded them of the swift frigate pursuit ships used by pirates in the tropical seas. Adult males are all black. Most birds seen in Hawai'i are females, identified by white throats and breast, or juveniles, with plumage like the female adult but with all-white heads.

Great Frigatebird ('Iwa), Fregata minor palmerstoni. (JIM DENNY)

Caspian Tern

STERNA CASPIA

Migrant
21 inches (53 cm)

This rare visitor is the largest tern in the world. It has a near-global distribution. The North American Pacific coast population has been increasing in recent years. Reports of the bird in Hawai'i have been on the increase as well. In the Hawaiian Islands, most sightings come from Maui and O'ahu, the two islands with the most expansive wetlands. The Caspian Tern has a large head and thick neck. The heavy red orange bill has a tinge of brown at the tip. The tail has a shallow fork. The bird is all white with the exception of a black cap and black outer primaries. Legs are black in adults and red in immature birds. The Caspian Tern feeds on fish that it obtains by picking from the surface or by diving headfirst into the water. When hunting, it cruises high above the water.

Caspian Tern, Sterna caspia. (JIM DENNY)

Common Tern

STERNA HIRUNDO

Migrant
14.5 inches (37 cm)

Contrary to the name, the Common Tern is not at all common in Hawai'i. It is in fact a rare visitor that is not reported every year. Most birds are seen in the winter months, and most are juveniles. This medium-sized tern is one of several that have been reported in Hawai'i. The similar-sized Arctic Tern *(Sterna paradisaea)* migrates through the Hawaiian Islands in large numbers in the spring but is almost entirely pelagic and rarely seen ashore. To identify any medium-sized tern seen on a Hawai'i reef or wetland, pay close attention to the color and length of the bill and legs. The Common Tern has longer legs and bill than the Arctic Tern and a distinctive dark carpal or shoulder bar. The smaller Least Tern *(Sterna antillarum)* is also a rare visitor to the main islands. Measuring only 9 inches (23 cm), it is pale gray above and all white below. The black-tipped bill and short legs are yellow. The crown and eye stripe are black.

Common Tern, Sterna hirundo. (JIM DENNY)

Gray-backed Tern *(Pākalakala)*

STERNA LUNATA

Indigenous
15 inches (38 cm)

This attractive tern is much more common in the Northwestern Hawaiian Islands, but it does nest in small numbers on Mokumanu, a small islet located off the windward coast of O'ahu. The sea near that breeding colony is the best place to look for this species. A few have been reported passing close to shore at the Hālona Blowhole on O'ahu. Most sightings have occurred between the months of April and October. It is also known as the Spectacled Tern or by its Hawaiian name, *Pākalakala*. It is gray above and white below with a long, forked gray tail. The crown is black, and a black line extends from the base of the bill through the eyes, giving it a spectacled appearance. The flight is more graceful than that of the more abundant Sooty Tern.

Gray-backed Tern (Pākalakala), Sterna lunata. (JACK JEFFREY)

Sooty Tern (*'Ewa'ewa*)

STERNA FUSCATA OAHUENSIS

Indigenous
17 inches (43 cm)

The Hawaiians named this species the *'Ewa'ewa.* Another appropriate name would be the "wide-awake tern" in reference to the incessant flight call it utters over island communities on misty, rainy nights. This graceful seabird is black above and all white below with a black malar stripe that extends to the base of the straight daggerlike bill. A roadside location where this bird can easily be seen with a spotting scope is the viewpoint near Makapu'u Point on the island of O'ahu. Look for them hovering over the small offshore island of Mānana. They nest there in great numbers between the months of February and September. The Sooty Tern is a very fast-flying bird. In flight, it resembles a small *'Iwa,* or Frigatebird.

The small island of Mānana lies just off the east coast of O'ahu near Sea Life Park. The island is an important breeding place for thousands of seabirds. A spotting scope is helpful when trying to identify the many birds seen over and around this small refuge. (JIM DENNY)

Sooty Tern ('Ewa'ewa), Sterna fuscata oahuensis.
(JACK JEFFREY)

Brown Noddy *(Noio kōhā)*

ANOUS STOLIDUS PILEATUS

Indigenous
17 inches (43 cm)

The Hawaiians named this species the *Noio kōhā*. It is also known as the Common Noddy. The name noddy is given to these birds because they constantly nod or bow to each other at breeding colonies. It is slightly larger than the Hawaiian Noddy and has a longer, darker tail. The flight is more purposeful and steady. The Brown Noddy feeds farther offshore than the more conspicuous Hawaiian Noddy. Thousands are known to nest on Lehua and Mokumanu islets. A look through a strong spotting scope near dusk from any headland on O'ahu or Kaua'i should find the bird. The nesting period of May to August is the best time to look for it.

Brown Noddy (Noio kōhā), Anous stolidus pileatus. (JACK JEFFREY)

Hawaiian Noddy *(Noio)*

ANOUS MINUTUS MELANOGENYS

Endemic
15 inches (38 cm)

The Hawaiian name for this species is *Noio*. It is also known as the Black Noddy or White-capped Noddy. Local fishermen often refer to it as the Pachi Pachi bird. The Hawaiian race, *melanogenys*, differs from Black Noddies of other locales by orange legs and a paler rump and tail. The *Noio* nests in small colonies on ledges and in cavities on precipitous rocky cliffs near the shore. They can usually be seen feeding year-round near the nest sites. Look for a dark gray seabird flying in an erratic pattern very close to the ocean surface with rapid flapping and very little gliding. Birds can easily be seen on boat tours along Kaua'i's Nāpali coast. The two best roadside locations are Wai'ānapanapa State Park on the island of Maui and the Hōlei Sea Arch on the Chain of Craters Road in Hawai'i Volcanoes National Park.

Hawaiian Noddy (Noio), Anous minutus melanogenys. (JACK JEFFREY)

White Tern (*Manu o Kū*)

GYGIS ALBA

Indigenous
12 inches (31 cm)

The White Tern is plentiful in the Northwestern Hawaiian Islands, but in the main islands it can easily be seen only on the south shore of O'ahu. It was first recorded breeding there in 1961. Good places to look are in the large trees at Fort DeRussy, Kapi'olani Park, 'Iolani Palace, Thomas Square, and along busy Kalākaua Avenue. If you sort through the many white Rock Pigeons, you can often find it perched on a large horizontal branch. It is pure white with a long, pointed black bill. A ring of black feathers around the eye gives it a wide-eyed innocent look. The White Tern is graceful in flight and sometimes noisy near the nest site. Oddly enough, this beautiful seabird does not construct a nest. It chooses to lay its egg on a bare branch.

White Tern (Manu o Kū), Gygis alba. (JACK JEFFREY)

APPENDIX 1
Species Distribution

A, abundant; **C**, common; **E,** species listed as "endangered" by the U.S. Fish and Wildlife Service; **R**, rare; **T**, "threatened"; **U**, uncommon

SPECIES	Kaua'i	O'ahu	Moloka'i	Lāna'i	Maui	Hawai'i
Urban Birds						
Rose-ringed Parakeet *Psittacula krameri*	C	C			R	R
Red-crowned Parrot *Amazona viridigenalis*		U				
Red-masked Parakeet *Aratinga erythrogenys*		U				U
Mitred Parakeet *Aratinga mitrata*					U	U
Patagonian Parakeet *Cyanoliseus patagonus*						R
Japanese White-eye (Mejiro) *Zosterops japonicus*	A	A	A	A	A	A
Spotted Dove *Streptopelia chinensis*	A	A	A	A	A	A
Rock Pigeon *Columba livia*	A	A	A	A	A	A
Zebra Dove *Geopelia striata*	A	A	A	A	A	A
Mourning Dove *Zenaida macroura*		R	R		U	R

SPECIES	Kaua'i	O'ahu	Moloka'i	Lāna'i	Maui	Hawai'i
Urban Birds, cont....						
Red-vented Bulbul *Pycnonotus cafer*		A				
Red-whiskered Bulbul *Pycnonotus jocosus*		A				
Common Myna *Acridotheres tristis*	A	A	A	A	A	A
Northern Mockingbird *Mimus polyglottos*	C	U	C	C	C	C
Northern Cardinal *Cardinalis cardinalis*	C	C	C	C	C	C
Red-crested Cardinal *Paroaria coronata*	C	C	C	C	C	
Yellow-billed Cardinal *Paroaria capitata*						C
House Sparrow *Passer domesticus*	A	A	A	A	A	A
House Finch *Carpodacus mexicanus*	A	A	A	A	A	A
Saffron Finch *Sicalis flaveola*	R	U			R	C
Yellow-fronted Canary *Serinus mozambicus*		C	C			C
Yellow-faced Grassquit *Tiaris olivacea*		U				
Red Avadavat *Amandava amandava*	C	C			R	C
Common Waxbill *Estrilda astrild*	U	A	R		R	U
Black-rumped Waxbill *Estrilda troglodytes*						R
Lavender Waxbill *Estrilda caerulescens*		R			R	U
Orange-cheeked Waxbill *Estrilda melpoda*		R			U	
Red-cheeked Cordonbleu *Uraeginthus bengalus*						R
Java Sparrow *Padda oryzivora*	C	C	R		C	C

SPECIES	Kaua'i	O'ahu	Moloka'i	Lāna'i	Maui	Hawai'i
Urban Birds, cont....						
African Silverbill *Lonchura cantans*	U	U	U	C	C	C
Chestnut Munia *Lonchura atricapilla*	A	C		R	U	R
Nutmeg Mannikin *Lonchura punctulata*	C	C	C	C	C	C
Country Birds						
Hawaiian Hawk *('Io)* E *Buteo solitarius*						C
Northern Harrier *Circus cyaneus*	R	R	R		R	
Osprey *Pandion haliaetus*	R	R	R	R	R	R
Peregrine Falcon *Falco peregrinus*	R	R	R		R	R
Hawaiian Owl *(Pueo)* *Asio flammeus sandwichensis*	C	R	U	U	C	C
Barn Owl *Tyto alba*	C	C	C	C	C	C
Cattle Egret *Bubulcus ibis*	A	C	C	C	C	C
Black Francolin *Francolinus francolinus*	C	R	C		U	C
Gray Francolin *Francolinus pondicerianus*	R	U	C	C	C	C
Erckel's Francolin *Francolinus erckelii*	C	U	R	C		C
Chukar *Alectoris chukar*	R	R	C	C	C	C
California Quail *Callipepla californica*	R		U	U	R	C
Gambel's Quail *Callipepla gambelii*				C		R
Japanese Quail *Coturnix japonica*	U	R	R	R	R	U
Wild Turkey *Meleagris gallopavo*	U	R	U	U	U	U

SPECIES	Kaua'i	O'ahu	Moloka'i	Lāna'i	Maui	Hawai'i
Country Birds, cont...						
Common Peafowl *Pavo cristatus*		R			R	R
Ring-necked Pheasant *Phasianus colchicus*	C	U	U	U	C	C
Green Pheasant *Phasianus colchicus versicolor*	R					R
Chestnut-bellied Sandgrouse *Pterocles exustus*						R
Sky Lark *Alauda arvensis*	R	U	U	R	C	C
Western Meadowlark *Sturnella neglecta*	C					
Forest Birds						
Hawai'i Crow *('Alalā)* E *Corvus hawaiensis*						R
Hawai'i *'Elepaio* *Chasiempis sandwichensis sandwichensis*						C
O'ahu *'Elepaio* E *Chasiempis sandwichensis ibidis*		R				
Kaua'i *'Elepaio* *Chasiempis sandwichensis sclateri*	C					
'Ōma'o *Myadestes obscurus*						C
Small Kaua'i Thrush E *(Puaiohi)* *Myadestes palmeri*	R					
Palila E *Loxioides bailleui*						R
'Akakane E *(Hawai'i 'Ākepa)* *Loxops coccineus coccineus*						R

SPECIES	Kaua'i	O'ahu	Moloka'i	Lāna'i	Maui	Hawai'i
Forest Birds, cont....						
'Akeke'e E *Loxops caeruleirostris*	R					
'Anianiau *Magumma parvus*	C					
Kaua'i *'Amakihi* *Hemignathus kauaiensis*	C					
O'ahu *'Amakihi* *Hemignathus flavus*		U				
Hawai'i *'Amakihi* *Hemignathus virens virens*						C
Maui *'Amakihi* *Hemignathus virens wilsoni*			U		C	
'Akiapōlā'au E *Hemignathus munroi*						R
Kaua'i Creeper E (*'Akikiki*) *Oreomystis bairdi*	R					
Hawai'i Creeper E *Oreomystis mana*						R
Maui Creeper (Maui *'Alauahio*) *Paroreomyza montana*					C	
Maui Parrotbill E *Pseudonester xanthophrys*					R	
'Apapane *Himatione sanguinea*	C	C	U	R	C	C
'I'iwi *Vestiaria coccinea*	U	R	R		C	C
Crested Honeycreeper E (*'Ākohekohe*) *Palmeria dolei*					U	
Melodious Laughing- thrush (Hwamei) *Garrulax canorus*	C	U	R		U	C
Greater Necklaced Laughingthrush *Garrulax pectoralis*	R					

SPECIES	Kaua'i	O'ahu	Moloka'i	Lāna'i	Maui	Hawai'i
Forest Birds, cont….						
Japanese Bush-Warbler (Uguisu) *Cettia diphone*	C	C	C	C	C	R
Red-billed Leiothrix *Leiothrix lutea*		C	U		C	C
Mariana Swiftlet *Aerodramus bartschi*		R				
Red Junglefowl *(Moa)* *Gallus gallus*	C					
Kalij Pheasant *Lophura leucomelanos*						C
Wetland Birds						
Pied-billed Grebe *Podilymbus podiceps*	R	R			R	R
Great Blue Heron *Ardea herodias*	R	R	R		R	R
Black-crowned Night-Heron *('Auku'u)* *Nycticorax nycticorax hoactli*	C	C	C	C	C	C
White-faced Ibis *Plegadis chihi*	R	R	R		R	R
Fulvous Whistling-Duck *Dendrocygna bicolor*	R	R	R		R	
Greater White-fronted Goose *Anser albifrons*	R	R	R		R	R
Snow Goose *Chen caerulescens*	R	R	R		R	R
Canada Goose *Branta canadensis*	R	R	R		R	R
Cackling Goose *Branta hutchinsii*	R	R			R	R
Hawaiian Goose *(Nēnē)* E *Branta sandvicensis*	U		R		U	U
Brant *Branta bernicla nigricans*	R	R	R		R	R
Tundra Swan *Cygnus columbianus*	R		R		R	
Gadwall *Anas strepera*		R	R		R	R

SPECIES	Kauaʻi	Oʻahu	Molokaʻi	Lānaʻi	Maui	Hawaiʻi
Wetland Birds, cont....						
Eurasian Wigeon *Anas penelope*	R	R	R		R	R
American Wigeon *Anas americana*	U	U	U		U	U
Mallard *Anas platyrhynchos*	R	R	R		R	R
Hawaiian Duck E (*Koloa maoli*) *Anas wyvilliana*	C	R	R		R	U
Blue-winged Teal *Anas discors*	R	R	R		R	R
Cinnamon Teal *Anas cyanoptera*	R	R	R		R	
Northern Shoveler (*Koloa mohā*) *Anas clypeata*	C	C	C	R	C	C
Northern Pintail (*Koloa māpu*) *Anas acuta*	C	C	C	R	C	C
Green-winged Teal *Anas crecca*	R	R	R		R	R
Eurasian Teal *Anas crecca crecca*	R	R	R		R	R
Canvasback *Aythya valisineria*	R	R	R		R	R
Redhead *Aythya americana*	R	R			R	R
Ring-necked Duck *Aythya collaris*	U	U	U		U	U
Tufted Duck *Aythya fuligula*	R	R			R	R
Greater Scaup *Aythya marila*	R	R	R		R	R
Lesser Scaup *Aythya affinis*	U	U	R		U	U
Bufflehead *Bucephala albeola*	U	U	R		U	R
Hooded Merganser *Lophodytes cucullatus*	R	R	R			R
Red-breasted Merganser *Mergus serrator*	R	R			R	R

SPECIES	Kaua‘i	O‘ahu	Moloka‘i	Lāna‘i	Maui	Hawai‘i
Wetland Birds, cont....						
Ruddy Duck *Oxyura jamaicensis*		R			R	R
Hawaiian Coot E (*‘Alae ke‘oke‘o*) *Fulica alai*	C	C	C	R	C	C
Hawaiian Moorhen E (*‘Alae ‘ula*) *Gallinula chloropus sandvicensis*	C	C				
Hawaiian Stilt (*Ae‘o*) E *Himantopus mexicanus knudseni*	C	C	C	R	C	U
Black-bellied Plover *Pluvialis squatarola*	U	U	U		U	U
Pacific Golden-Plover (*Kōlea*) *Pluvialis fulva*	C	C	C	C	C	C
Semipalmated Plover *Charadrius semipalmatus*	U	U	U		U	U
Killdeer *Charadrius vociferus*	R	R			R	R
Greater Yellowlegs *Tringa melanoleuca*	R	R			R	
Lesser Yellowlegs *Tringa flavipes*	U	U	R		U	U
Wandering Tattler (*‘Ūlili*) *Heteroscelus incanus*	C	C	C	C	C	C
Spotted Sandpiper *Actitis macularia*	R	R			R	R
Whimbrel *Numenius phaeopus*		R	R		R	R
Bristle-thighed Curlew (*Kioea*) Numenius tahitiensis	R	U	U		R	R
Marbled Godwit *Limosa fedoa*		R			R	R
Bar-tailed Godwit *Limosa lapponica*	R	R			R	R
Ruddy Turnstone (*‘Akekeke*) *Arenaria interpres*	C	C	C	C	C	C
Red Knot *Calidris canutus*		R	R		R	R
Sanderling (*Hunakai*) *Calidris alba*	C	C	C	C	C	C

SPECIES	Kaua'i	O'ahu	Moloka'i	Lāna'i	Maui	Hawai'i
Wetland Birds, cont....						
Semipalmated Sandpiper *Calidris pusilla*	R	R	R		R	
Western Sandpiper *Calidris mauri*	R	R	R		R	
Red-necked Stint *Calidris ruficollis*		R			R	
Least Sandpiper *Calidris minutilla*	U	U	R		U	U
Pectoral Sandpiper *Calidris melanotos*	U	U	U		U	U
Sharp-tailed Sandpiper *Calidris acuminata*	U	U	U	U	U	U
Dunlin *Calidris alpina*	U	U	U		U	U
Curlew Sandpiper *Calidris ferruginea*		R			R	
Ruff *Philomachus pugnax*	R	R	R		R	R
Short-billed Dowitcher *Limnodromus griseus*		R			R	
Long-billed Dowitcher *Limnodromus scolopaceus*	U	U	U		U	U
Wilson's Snipe *Gallinago delicata*	R	R	R		R	R
Wilson's Phalarope *Phalaropus tricolor*	R	R	R		R	R
Red Phalarope *Phalaropus fulicarius*	R	R			R	R
Pomarine Jaeger *Stercorarius pomarinus*	R	R		R	R	R
Laughing Gull *Leucophaeus atricilla*	U	U	U		U	U
Bonaparte's Gull *Chroicocephalus philadelphia*	R	R			R	R
Ring-billed Gull *Larus delawarensis*	U	U	U		U	U
Glaucous-winged Gull *Larus glaucescens*	U	U	U	U	U	U

SPECIES	Kaua'i	O'ahu	Moloka'i	Lāna'i	Maui	Hawai'i
Wetland Birds, cont....						
Glaucous Gull *Larus hyperboreus*	R	R				R
Franklin Gull *Larus pipixcan*	R	R	R		R	R
California Gull *Larus californicus*		R			R	
Herring Gull *Larus argentatus*		R			R	
Western Gull *Larus occidentalis*		R				
Seabirds						
Laysan Albatross *(Mōlī)* *Phoebastria immutabilis*	C	U	R	R	R	R
Black-footed Albatross *Phoebastria nigripes*	R	R	R	R	R	R
Hawaiian Petrel *('Ua'u)* E *Pterodroma sandwichensis*	R	R	R	R	R	R
Bulwer's Petrel *('Ou)* *Bulweria bulwerii*	R	R	R	R	R	R
Wedge-tailed Shearwater *('Ua'u kani)* *Puffinus pacificus chlororhynchus*	C	U	R	R	U	U
Christmas Shearwater *Puffinus nativitatis*	R	R				
Newell's Shearwater *('A'o)* *Puffinus auricularis newelli*	U		R			R
Band-rumped Storm-Petrel *('Akē'akē)* *Oceanodroma castro*	R				R	R
White-tailed Tropicbird *(Koa'e kea)* *Phaethon lepturus dorotheae*	C	C	C	C	U	C

SPECIES	Kaua'i	O'ahu	Moloka'i	Lāna'i	Maui	Hawai'i
Seabirds, cont....						
Red-billed Tropicbird *Phaethon aethereus*	R	R				
Red-tailed Tropicbird *(Koa'e 'ula) Phaethon rubricauda*	C	C	U	U	U	U
Masked Booby *('Ā)* *Sula dactylatra personata*	R	U				R
Brown Booby *('Ā)* *Sula leucogaster plotus*	C	C	C	C	C	C
Red-footed Booby *('Ā)* *Sula sula rubripes*	C	C	U	U	U	U
Great Frigatebird *('Iwa)* *Fregata minor palmerstoni*	C	C	U	U	U	U
Caspian Tern *Sterna caspia*		R			R	
Common Tern *Sterna hirundo*	R	R			R	R
Arctic Tern *Sterna paradisaea*	R	R			R	R
Least Tern *Sterna antillarum*		R			R	
Gray-backed Tern *(Pākalakala) Sterna lunata*	R	R				
Sooty Tern *('Ewa'ewa)* *Sterna fuscata oahuensis*	R	A			R	R
Brown Noddy *(Noio kōhā)* *Anous stolidus pileatus*	U	U				
Hawaiian Noddy *(Noio)* *Anous minutus melanogenys*	C	U	R	R	C	C
White Tern *(Manu o Kū)* *Gygis alba*		C				R

APPENDIX 2

Suggested Birding Locations

KAUA'I

Alaka'i Swamp Trail
Hanalei National Wildlife Refuge
Hanapēpē salt pond and airfield
Hulē'ia Stream Valley
Kalalau Valley scenic viewpoints
Kawai'ele Sand Mine
Kīlauea pastures
Kīlauea Point National Wildlife Refuge
Kukuiolono Park
Kukui'ula Boat Harbor and adjacent wetlands
Līhu'e Airport
Mānā plains
Mōhihi-Wai'alae Trail
Nāpali Coast
Pihea Trail
Waimea Canyon Road

LĀNA'I

Hulopo'e Beach Park
Keōmuku Road

O'AHU

'Aiea Loop Trail
Fort DeRussy
Foster Botanic Garden
Hālona Blowhole
Honolulu Zoo

O'AHU, CONT....

Ho'omaluhia Botanical Garden
Ka'ena Point
Kahuku wetlands
Kapi'olani Park
Kawainui Marsh and Ka'elepulu Stream
Kualoa Regional Park
Kuli'ou'ou Valley Trail
Lyon Arboretum
Mā'ili Beach Park
Makapu'u Point and Mānana Island
Māmala Bay
Round Top Drive
Sand Island Park
Wa'ahila Ridge State Recreation Area
West Loch Shoreline Park

MOLOKA'I

Duke Maliu Regional Park
Kakahai'a Beach Park
Kapuāiwa Coconut Grove
Kōheo coastal wetland
Maui Education Center lawn
Pāpōhaku Beach Park

MAUI

Central Maui plains
Haleakalā National Park
Hosmer Grove

MAUI, CONT....

ʻĪao Stream Park
Kanahā Pond State Wildlife Sanctuary
Keālia National Wildlife Refuge
Keʻanae wetlands
Kula Botanical Gardens
Polipoli State Recreation Area
Waiʻānapanapa Caves State Park
Waikamoi Preserve*

BIG ISLAND

ʻAimakapā Pond
Big Island Country Club
Hakalau National Wildlife Refuge*
Hawaiʻi Volcanoes National Park
Hilo ponds
Hōlei Sea Arch
Kailua-Kona hotel grounds and parks
Keauhou Small Boat Harbor
Kīpuka Puaʻulu (Bird Park)
Kīpuka 21 (mile marker 21 of Saddle Road)
Manukā State Park
Mauna Kea Access Road
Mauna Kea State Park (Pōhakuloa)
Puʻu Anahulu
Puʻu Laʻau
Puʻu ʻŌʻō Trail
Waimea plains

*Special permission for access is required.

For detailed directions to many of these locations and others, the author highly recommends four references:

Enjoying Birds in Hawaii: A Birdfinding Guide to the Fiftieth State. 3rd ed., by H. Douglas Pratt. Honolulu: Mutual Publishing Company, 2003.

The Birdwatcher's Guide to Hawaiʻi, by Rick Soehren. Honolulu: University of Hawaiʻi Press, 1996.

Birding Hawaii. A Web site by Christian W. Melgar. http://www.birdinghawaii.co.uk.

Reference Maps of the Islands of Hawaiʻi: Kauaʻi, Oʻahu, Molokaʻi and Lānaʻi, Maui, and Hawaiʻi, by James A. Bier. Honolulu: University of Hawaiʻi Press.

REFERENCES

Anonymous. Mrs. Isenberg Aids Movement to Get Birds. *Honolulu Advertiser*, 19 March 1930, p. 9.

———. Shipment of Rare Birds Coming Today. *Honolulu Advertiser*, 3 December 1930, p. 4.

———. Board Permits Rare Birds to Enter at Last. *Honolulu Advertiser*, 6 December 1935, p. 9.

Atkinson, Carter T., Robert J. Dusek, and William M. Iko. Avian Malaria Fatal to Juvenile ʻIʻiwi. *Hawaii's Forest and Wildlife* 8 (3): 1. 1993.

Berger, Andrew J. *Hawaiian Birdlife*. Honolulu: University of Hawaiʻi Press, 1981.

Bier, James A. Reference Maps of the Islands of Hawaiʻi: Hawaiʻi. 7th ed. Honolulu: University of Hawaiʻi Press, 2002.

———. Reference Maps of the Islands of Hawaiʻi: Molokaʻi and Lānaʻi. 5th ed. Honolulu: University of Hawaiʻi Press, 2002.

———. Reference Maps of the Islands of Hawaiʻi: Kauaiʻi. 7th ed. Honolulu: University of Hawaiʻi Press, 2004.

———. Reference Maps of the Islands of Hawaiʻi: Maui. 8th ed. Honolulu: University of Hawaiʻi Press, 2007.

———. Reference Maps of the Islands of Hawaiʻi: Oʻahu. 7th ed. Honolulu: University of Hawaiʻi Press, 2007.

Carlquist, Sherwin. *Hawaii: A Natural History*. Garden City, New York: Natural History Press, 1970.

Culliney, John L. *Islands in a Far Sea: Nature and Man in Hawaii*. San Francisco: Sierra Club Books, 1988.

Denny, Jim. *The Birds of Kauaʻi*. Honolulu: University of Hawaiʻi Press, 1999.

Freed, Leonard A., Sheila Conant, and Robert C. Fleischer. Evolutionary Ecology and Radiation of Hawaiian Passerine Birds. Page 335 *in* E. Alison Kay, ed. *A Natural History of the Hawaiian Islands: Selected Readings*. 2nd ed. Honolulu: University of Hawaiʻi Press, 1994.

Grant, P. J. *Gulls: A Guide to Identification*. San Diego: Academic Press, 1997.

Harrison, Craig S. *Seabirds of Hawaii: Natural History and Conservation*. Ithaca, New York: Comstock Publishing Associates, 1990.

Hawaiʻi Audubon Society. *Hawaii's Birds*. 6th ed. Honolulu: Hawaiʻi Audubon Society, 2005.

Melgar, Christian W. *Birding Hawaii*. http://www.birdinghawaii.co.uk (accessed 19 May 2008).

National Geographic Society. *Complete Birds of North America*. Washington: The National Geographic Society, 2006.

Olson, Storrs L. Bird/Plant Interactions in Hawai'i as Seen through the Fossil Record. Address presented at Kaua'i Community College, Līhu'e, 30 March 2007.

Perkins, R. C. L. *Fauna Hawaiiensis*. Cambridge, England: The University Press, 1903.

Pratt, H. Douglas. *Enjoying Birds in Hawaii: A Birdfinding Guide to the Fiftieth State*. 3rd ed. Honolulu: Mutual Publishing Company, 2003.

———. *The Hawaiian Honeycreepers*. New York: Oxford University Press, 2005.

Pratt, H. Douglas, Phillip L. Bruner, and Delwyn G. Berrett. *A Field Guide to the Birds of Hawaii and the Tropical Pacific*. Princeton: Princeton University Press, 1987.

Pratt, H. Douglas, and Jack Jeffrey. *A Pocket Guide to Hawai'i's Birds*. Honolulu: Mutual Publishing, 1996.

Pukui, Mary Kawena, and Samuel H. Elbert. *Hawaiian Dictionary*. Revised and enlarged ed. Honolulu: University of Hawai'i Press, 1986.

Pukui, Mary Kawena, Samuel H. Elbert, and Esther T. Mookini. *Place Names of Hawaii*. Revised and expanded ed. Honolulu: The University Press of Hawai'i, 1974.

Pyle, Robert L. Checklist of the Birds of Hawaii – 2002. *'Elepaio* 62:137–148, 2002.

Scott, Michael J., S. Mountainspring, F. L. Ramsey, and C. B. Kepler. *Forest Bird Communities of the Hawaiian Islands: Their Dynamics, Ecology, and Conservation*. Studies in Avian Biology, No. 9., 1986.

Scott, J. M., S. Conant, and C. Van Riper III, eds. Evolution, Ecology, Conservation and Management of Hawaiian Birds: A Vanishing Avifauna. Studies in Avian Biology, No. 22. 2001.

Sibley, David Allen. *The Sibley Guide to Birds*. National Audubon Society. New York: Chanticleer Press, 2000.

Soehren, Rick. *The Birdwatcher's Guide to Hawai'i*. Honolulu: University of Hawai'i Press, 1996.

State of Hawai'i. Department of Land and Natural Resources. Hawaii's CWCS. http://www.state.hi.us/dlnr/dofaw/cwcs (accessed 24 June 2008).

Stokes, Donald W., and Lillian Q. Stokes. *Stokes Field Guide to Birds, Western Region*. Boston: Little Brown and Company, 1996.

Wagner, Warren L., Derral R. Herbst, and S. H. Sohmer. *Manual of the Flowering Plants of Hawai'i*. 2 vols. Honolulu: University of Hawai'i Press and Bishop Museum Press, 1990.

Waring, Dr. George H. *Free-Ranging Parrot Population of Haiku Distict, Maui, Hawaii*. http://www.hear.org/alienspeciesinhawaii/waringreports/parrot.htm. (accessed 18 August 2008).

Williams, Richard N. *Bulbul Introductions on Oahu*. *'Elepaio* 43:89–90, 1983.

Ziegler, Alan C. Hawaiian Natural History, Ecology, and Evolution. Honolulu: University of Hawai'i Press, 2002.

INDEX